Results and Problems in Cell Differentiation

Series Editor:
W. Hennig

35

Springer
Berlin
Heidelberg
New York
Barcelona
Hong Kong
London
Milan
Paris
Tokyo

Karsten Weis (Ed.)

Nuclear Transport

With 32 Figures

Springer

Dr. KARSTEN WEIS
University of California, Berkeley
Department of Molecular & Cell Biology
Berkeley, CA 94720-3200
USA

ISSN 0080-1844
ISBN 3-540-42368-0 Springer-Verlag Berlin Heidelberg New York

Library of Congress Cataloging-in-Publication Data

Springer-Verlag Berlin Heidelberg New York
a member of BertelsmannSpringer Science+Business Media GmbH

http://www.springer.de

Production: PRO EDIT GmbH, Heidelberg, Germany
Cover concept: Meta Design, Berlin, Germany
Cover Production: design & production, Heidelberg, Germany
Typesetting: Best-set Typesetter Ltd., Hong Kong
Printed on acid-free paper SPIN 10765123 39/3130/Di 5 4 3 2 1 0

Preface

The hallmark of a eukaryotic cell is the nucleus. The evolution of a nucleus and the spatial separation of the genetic information from protein biosynthesis have led to the development of remarkable strategies to control gene expression in eukaryotes. However, it also necessitated the evolution of a complex machinery that transports macromolecules between the cytoplasm and the nucleus. Both proteins and RNAs must be imported into and exported out of the nucleus. The enormous number of transport events occur through large multiprotein complexes, which are embedded in the nuclear envelope and termed nuclear pore complexes (NPCs). The transport cargoes themselves are recognized by soluble receptors, which shuttle back and forth between the two compartments and in doing so ferry their load to the correct destination.

Research on the various nucleocytoplasmic transport pathways has made tremendous progress over the last couple of years. This book tries to provide a comprehensive overview of our current knowledge in this area of research. It is also intended to introduce scientists to this exciting field. The book contains nine chapters. The first two chapters summarize what we know about the structure and function of the nuclear pore complex (NPC). The work on the yeast *Saccharomyces cerevisiae* has led to a first assessment of the total protein composition of the NPC (Rout et al. 2000). Based on its size, it was surprising that the NPC probably contains less than 50 different proteins (albeit many are present in multiple copies). Another interesting observation was that many nuclear pore proteins (also called nucleoporins or Nups) can be found on both the cytoplasmic and nuclear side of the NPC. Strambio-de-Castilia and Rout review the structure and the composition of the yeast NPC in the first chapter. Despite its similar function, the vertebrate nuclear pore complex is bigger and its composition seems to differ significantly from the yeast NPC. Fahrenkrog and Aebi discuss these differences and describe the structure and the function of the vertebrate NPC.

A key regulator of nucleocytoplasmic transport is the small GTPase Ran. Bischoff et al. discuss the function and regulation of Ran. Ran is highly enriched in the nucleus and is pivotal to conferring directionality to many transport events since it regulates cargo binding to and cargo release from soluble transport receptors. Recently, it has been shown that Ran also plays an important role during mitosis and a function for Ran in mitotic spindle assembly and nuclear envelope formation has been demonstrated (reviewed in Dasso 2001). A model has emerged that views Ran as a positional marker defining the space around chromatin, which corresponds to the nucleus in interphase.

The multitude of transport pathways that operate between the cytoplasm and the nucleus seem to rely on the existence of an equally large protein family of soluble transport receptors. These receptors have been termed either importins and exportins or alternatively karyopherins (Kaps in yeast). In yeast, this protein family consists of 14 members and the latest count demonstrates the existence of probably more than 25 members in metazoans (reviewed in Ström and Weis 2001). For many of these transport receptors, transport substrates have now been identified. Interestingly, many of the nuclear transport factors characterized in yeast are not essential for viability, but they transport essential cargoes. This phenomenon can be best explained by the fact that cargoes may access alternative transport pathways. This is exemplified in the import pathway of ribosomal proteins, which is mediated by different import receptors (reviewed in Ström and Weis 2001).

A detailed picture has emerged as to how importins and exportins fulfill their function in nuclear transport. Importins bind to their cargoes in the cytoplasm and ferry them into the nucleus. In the nucleus, substrate release is induced through RanGTP, which is highly enriched in this compartment. In contrast, exportins bind to their nuclear cargoes only in the presence of RanGTP. After translocation to the cytoplasm, substrates are released through GTP hydrolysis by Ran completing the transport cycle. Fornerod and Ohno concentrate in their contribution on the function of the exportins that mediate the export of proteins and ribonucleoprotein complexes. Another exportin-mediated transport pathway is discussed in the chapter by Simos et al., who review the nuclear export of tRNA by Los1/exportin-t.

A wealth of structural information on nuclear transport factors has recently become available. Crystal structures of different import receptors complexed with either substrates or Ran have been solved. Conti reviews these findings and discusses their implication for our understanding of the mechanism of nuclear transport.

Although nuclear export through members of the exportin family is a major path, it does not seem to be the only way to exit the nucleus. Izaurralde discusses the export of messenger RNA. Work on both yeast and metazoans has altered our view of this very important and conserved step of gene expression, thus providing a first glimpse of how mRNA export is regulated.

Viruses have been an extremely useful tool for many scientists working in the field of cell biology. This is particularly true for nuclear transport. For example, the first nuclear import and nuclear export signals were defined and mapped in viral proteins (the NLS of the SV40 large T antigen and the NES of the HIV-1 Rev protein, respectively). Cullen illustrates what we have learned from the work in retroviruses with regards to nuclear mRNA and protein export. In the last chapter, on the basis of several well-studied examples, Schüller and Ruis discuss the different strategies that are employed to regulate nuclear transport. It becomes evident that the possibility of rapidly regulating the activity of proteins by changing their location was an important factor for the evolutionary success of eukaryotes and the development of metazoan organisms.

I want to take this opportunity to thank again all the authors who contributed to this book. Unfortunately, it was not possible to include all aspects of nuclear transport in one volume, but I hope that the conscientious efforts of all contributors will make this volume a useful reference book for a broad spectrum of scientists.

References

Dasso M (2001) Running on Ran: nuclear transport and the mitotic spindle. Cell 104:321–324

Rout MP, Aitchison JD, Suprapto A, Hjertaas K, Zhao Y, Chait BT (2000) The yeast nuclear pore complex: composition, architecture, and transport mechanism. J Cell Biol 148:635–651

Ström AC, Weis K (2001) Importin-beta-like nuclear transport receptors. Genome Biol 2:REVIEWS3008

Karsten Weis
Berkeley, CA, USA, August, 2001

Contents

Using Retroviruses To Study the Nuclear Export of mRNA
Bryan R. Cullen

Regulated Nuclear Transport
Christoph Schüller and Helmut Ruis

The Structure and Composition of the Yeast NPC

Caterina Strambio-de-Castillia[1] and Michael P. Rout[1]

1 Introduction

The double-membraned nuclear envelope (NE) behaves as a selective barrier that segregates the genome from all cytosolic processes. A highly regulated exchange system between these two compartments is essential for proper cell growth, progression through the cell cycle, accurate responses to developmental and extracellular signals and to maintain the functional integrity of the nucleus. The sole mediators of controlled nucleocytoplasmic transport are the nuclear pore complexes (NPCs), large proteinaceous machineries embedded within specialized circular pores that traverse the NE. In actively growing cells it is estimated that every minute hundreds of proteins and ribonucleoprotein particles (RNPs) traverse each NPC in both directions. The basic mechanisms of nuclear transport appear to be highly conserved across distantly related species (reviewed in Nigg 1997; Mattaj and Englmeier 1998; Gorlich and Kutay 1999; Wente 2000). Although metabolites, water, ions and small macromolecules can freely diffuse through aqueous channels of 10 nm in the NPC, large macromolecular particles with a diameter of up to 30 nm are selectively transported across the NPC via a highly regulated energy-dependent process. Active transport requires specific soluble transport factors that recognize individual substrates both inside and outside the nucleus and mediate their interaction with the stationary phase of the NPC translocation machinery. Specifically, the translocation of transport substrates is known to require the docking of the transport complex to the NPC, the active translocation of the docked complexes across the NPC and the release of the substrate into the target compartment. Various models have been proposed to explain how this docked complex is actively translocated across the 50–60 nm long NPC transporter and then subsequently released into the nucleoplasm, and the matter is still highly controversial (see below). All models agree in attributing a crucial importance to the protein Ran in maintaining vectorial cargo transport and regulating the binding and release steps that take place during translocation. As a member of the Ras superfamily of small GTPases, Ran exists within the cell in a GDP-

[1] The Laboratory of Cellular and Structural Biology, The Rockefeller University, 1230 York Ave, Box 213, New York, NY 10021, USA

Results and Problems in Cell Differentiation, Vol. 35
K. Weis (Ed.): Nuclear Transport

bound and in a GTP-bound form. The balance between these two forms is regulated by a variety of Ran cofactors that are asymmetrically distributed within the cell. As a consequence cytoplasmic Ran is thought to exist prevalently in the GDP-bound form, while Ran-GTP is thought to predominate in the nucleus. This differential distribution of Ran-GTP versus Ran-GDP would establish directional transport by ensuring that transport complexes are formed in one compartment and disassembled in the other (reviewed in Cole and Hammell 1998; Mattaj and Englmeier 1998; Pemberton et al. 1998; Wozniak et al. 1998; Gorlich and Kutay 1999). Understanding this regulated transport demands an understanding of the detailed three-dimensional map of the NPC and of the interactions and relationships between the soluble and stationary phases of nuclear transport.

NPCs are present in all eukaryotic cells, and despite interesting differences in details, their morphology is remarkably conserved among evolutionary divergent phyla (Maul 1977; Yang et al. 1998). This makes studies using model organisms relevant to all eukaryotes. In particular, several characteristics make the yeast *Saccharomyces cerevisiae* an excellent model system to investigate the structure and composition of NPCs, and their role in nucleocytoplasmic transport. *S. cerevisiae* undergoes a closed mitosis such that the NE remains intact during cell division allowing the isolation and study of NPC components from all stages of the cell cycle. Furthermore, like many other aspects of yeast cell biology, the process of nucleocytoplasmic exchange appears to lack many of the complicated elaborations present in metazoans rendering it a much simpler model system. The genetics and molecular biology of yeast are better understood than those of any other eukaryote, and the genome is fully sequenced greatly speeding the process of identification and characterization of unknown proteins. Finally, many sophisticated cell biological and biochemical analysis techniques have been developed for this organism.

2 Overview of the Yeast NPC Structure

NPCs from all eukaryotes share a common architecture, and many NPC proteins (collectively named nucleoporins or nups) are conserved across phyla. In metazoans, the core of the NPC consists of a cylinder with a plane of pseudo-mirror symmetry running parallel to the NE, composed of eight interconnecting spoke-like structures symmetrically arranged around a "central transporter" (or central channel; Unwin and Milligan 1982; Hinshaw et al. 1992; Akey and Radermacher 1993). Electron microscopic images of the central transporter suggest that it is a centrally tapered hollow tube that spans the entire width of the NPC (Akey 1990; Akey and Radermacher 1993; Goldberg and Allen 1996; Kiseleva et al. 1998). Functional studies employing colloidal gold particles attached to nuclear-targeted proteins indicate that the NPC has a central hole with a functional diameter of ~9 nm that allows the free diffusion of small molecules but restricts passive diffusion of macromolecules

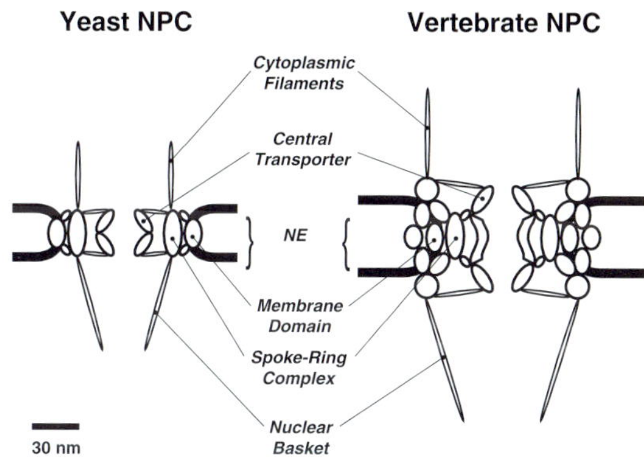

Fig. 1. Scale diagram of an idealized vertical section from a yeast NPC and a vertebrate NPC. Protein domains visualized in electron micrographs are shown as *ovals*. (Adapted from Yang et al. 1998)

through the pores. However, during active transport the central transporter can accommodate the passage of massive substrates, such as ribosomal subunits and pre-mRNPs, with a diameter up to 26 nm (Paine et al. 1975; Feldherr et al. 1984; Dworetzky and Feldherr 1988; Feldherr and Akin 1994a,b, 1997; Feldherr et al. 1998; Kiseleva et al. 1998). Peripherally associated nuclear and cytoplasmic filaments project from the core and are distinctly asymmetrical. Although the cytoplasmic filaments spread outwards perpendicularly to the central plane of the NE, the nuclear filaments conjoin to form the "nuclear fishtrap" or "nuclear basket" (Ris 1991; Goldberg and Allen 1992). Transport substrates dock to these peripheral filaments and translocate through the transporter on their way in and out of the nucleus.

The NPCs of *Saccharomyces* share many common features with their vertebrates counterparts, although they are significantly smaller both in mass and in volume (Fig. 1; Fahrenkrog et al. 1998; Yang et al. 1998). Interestingly, the differences in NPC size and mass between yeast and vertebrates can be accounted for by a concomitant simplification of the structure. In substance, the data are consistent with the hypothesis that the yeast NPC comprises only the central core of the vertebrate NPC and lacks many of the peripheral attachments including the lumenal spoke ring, the nuclear ring, and the cytoplasmic ring with its attached cytoplasmic particles. Likewise, the cytoplasmic fibers and nuclear basket appear to be conserved but are anchored to more central domains of the spoke-ring assembly. The central transporter is also smaller and appears to be missing a central cylinder, which gives the vertebrate transporter its hour-glass shape (Rout and Blobel 1993; Yang et al. 1998). The results of these comparative studies suggest that the architecture of an NPC can vary considerably and still be functional. Thus, the yeast NPC is likely to have retained or recapitulated the features that characterize what a streamlined NPC might look like. Accordingly, the yeast NPC is able to ensure the efficient exchange of material between the nucleus and the cytoplasm but apparently lacks the higher order structures necessary in multi-cellular organisms.

3 Yeast Nucleoporins: What's NUP, What's Not

A variety of immunological, biochemical and genetic techniques have been successfully employed in the past few years to identify yeast NPC components (reviewed in Rout and Wente 1994; Doye and Hurt 1997; Fabre and Hurt 1997). Of course, given the highly dynamic nature both of NPCs and of their interactions (see below), it is not always possible to establish what constitutes a "complete NPC". On first approximation, this problem can be generally resolved by adopting an operational definition such as considering bona fide nucleoporins to be those proteins that are stably associated with the NPC. Thus, the candidate nucleoporin should immunolocalize to the NPC by immunofluorescence (IF) microscopy or better by immunoelectron microscopy (IEM), should cofractionate with the NPC in subcellular fractionation procedures, and should interact genetically and biochemically with other known nucleoporins.

3.1 General Characteristics of Yeast Nucleoporins

Recent work has allowed us to set an upper limit for the total number of nucleoporins in yeast, and establish a rough map of the distribution of all known nucleoporins in the context of the three-dimensional map of the yeast NPC (Kraemer et al. 1995; Nehrbass et al. 1996; Fahrenkrog et al. 1998; Hurwitz et al. 1998; Marelli et al. 1998; Kosova et al. 1999; Strahm et al. 1999; Bailer et al. 2000; Rout et al. 2000). The total number of bona fide yeast nucleoporins is now estimated to be ~30 (Rout et al. 2000). This is a surprisingly low number for such a massive structure, especially considering that, for example, the much smaller ribosome is composed of ~80 different proteins. This apparent discrepancy between size and composition can be resolved by considering the high level of symmetry displayed by NPCs. First, most nucleoporins are symmetrically distributed with respect to the central plane of the NE. Hence the majority of yeast nucleoporins are present in two to four copies per spoke and are therefore present in 16 to 32 copies per NPC (Rout et al. 2000). Thus, the yeast NPC appears to be composed mainly of 16 copies of a subset of nucleoporins: eight copies facing the nucleus and eight copies facing the cytoplasm. If one considers that the average molecular weight of individual yeast nucleoporins is relatively high (~100 kDa), one can calculate that ~30 proteins each present in an average of 16 copies would produce a structure of ~50 MDa, thus completely accounting for the mass of the yeast NPC (measured to be between 55 and 66 MDa; Rout and Blobel 1993; Yang et al. 1998). At least 65% of yeast

S. cerevisiae nucleoporins have direct orthologs in vertebrate genomes attesting once again to the validity of this as a model system for the study of NPCs and in general nucleocytoplasmic transport processes. Yeast nucleoporins can be divided into three partially overlapping classed based on their sequence characteristics and presumed function: FG nucleoporins, non-FG nucleoporins, and pore membrane proteins (POMs).

3.1.1 FG Nucleoporins

Nearly half (12 out of 30) of yeast nucleoporins belong to the "FG nucleoporins" family (reviewed in Rout and Wente 1994; Fabre et al. 1995). These are characterized by the presence of at least one domain containing multiple GLFG, FXFG, or FG amino acid repeat motifs separated by polar spacer sequences, and are generally thought to be filamentous in nature (Buss et al. 1994). The spacer sequences between the FXFG and the FG repeats are generally highly charged and rich in serine and threonine residues. The GLFG spacers are generally devoid of acidic residues and have a prevalence of asparagine and glutamine residues. The role of these proteins in NPC translocation has been firmly established on the basis of numerous biochemical and genetic analyses (reviewed in Doye and Hurt 1997; Fabre and Hurt 1997; Ohno et al. 1998; Ryan and Wente 2000).

FG nucleoporins are known to provide the NPC docking sites for soluble transport factors associated to their cognate transport substrates. This was convincingly demonstrated using several methods. In vitro experiments performed using purified components have demonstrated that FG repeat motifs of various nucleoporins interact directly with members of the β-karyopherin family of transport factors and with other soluble transport factors (Rexach and Blobel 1995; Nehrbass and Blobel 1996). In addition, ex vivo biochemical studies have shown that β-karyopherins and other non-karyopherin transport factors interact specifically with FG nucleoporins or with truncated forms of these nucleoporins containing FG repeat motifs (Radu et al. 1995; Aitchison et al. 1996; Iovine and Wente 1997; Pemberton et al. 1997; Rout et al. 1997; Katahira et al. 1999; Hurt et al. 2000). Finally, in vivo localization analyses of various reporter transport substrates in yeast strains carrying truncated or otherwise non-functional mutant forms of various FG nucleoporins have demonstrated the physiological relevance of these docking interactions to ensure appropriately regulated nucleocytoplasmic transport (reviewed in Doye and Hurt 1997; Fabre and Hurt 1997).

Although the NPCs contain an extremely high number of FG repeat motifs and there is considerable overlap between different transport factors for their ability to bind certain FG nucleoporins, individual transport factors have strong preferences for specific docking sites at the NPC (Rexach and Blobel 1995; Aitchison et al. 1996; Rout et al. 1997; Marelli et al. 1998; Floer and Blobel 1999). This might indicate that there are different transport routes across the

NPCs which are utilized at the same time by different transport factors en route to their final destination.

Finally, the binding affinities of nuclear transport factors to their FG docking sites on the NPC are modulated by the nucleotide-bound status of Ran (Rexach and Blobel 1995; Floer and Blobel 1996, 1999; Nehrbass and Blobel 1996; Floer et al. 1997; Hood and Silver 1998; Solsbacher et al. 1998; Bayliss et al. 1999; Seedorf et al. 1999). Generally speaking, Ran-GTP is more abundant in the nucleus whereas Ran-GDP is prevalently cytoplasmic (see above). Consistent with this observation, nuclear import factors are dependent on Ran-GDP for binding to the NPC and are dissociated from the NPC by Ran-GTP. On the other hand, the affinity of transport factors carrying export substrates to the cytoplasm for specific FG nucleoporins is increased by Ran-GTP and reduced by Ran-GDP.

3.1.2 Non-FG Nucleoporins

The nucleoporins not containing obvious FG repeat sequences (collectively referred to as the non-FG nucleoporins) together are estimated to comprise ~70% of the total mass of the NPC (excluding POMs; Rout et al. 2000). Only one-third of these proteins are essential, likely attesting to high number of contacts that each protein makes with its numerous neighbors; this group of proteins appear to give rise to a region of the NPC formed by a tight network of proteins, so that the removal of one component is unlikely to have a catastrophic effect. As expected, several of these proteins are either genetically or physically linked to one another (Aitchison et al. 1995b; Nehrbass et al. 1996; Zabel et al. 1996; Tcheperegine et al. 1999). In addition, they are characterized by a symmetrical distribution close to the middle plane of the NPC (Wozniak et al. 1994; Nehrbass et al. 1996; Rout et al. 2000). Such findings have led to the suggestion that these nucleoporins may be part of the structural core of the NPC (Doye et al. 1994; Aitchison et al. 1995a,b; Li et al. 1995; Goldstein et al. 1996; Nehrbass et al. 1996; Siniossoglou et al. 1996; Zabel et al. 1996). The observation that Nic96p and Nup170p are physically linked to FG nucleoporins, and may contribute to their recruitment to the NPC (Grandi et al. 1993, 1995b; Kenna et al. 1996; Marelli et al. 1998), underscores this hypothesis and strongly suggests that that the NPC-core may constitute a foundation, on which more peripherally located nucleoporins involved in docking to nuclear transport factors are anchored and correctly positioned. Finally, by employing an in vivo assay to follow the free diffusion of various substrates across the NPC, Goldfarb and coworkers have recently provided new evidence that strongly suggests a role for two NPC core components, Nup170p and Nup188p, in determining the functional diameter of the passive diffusion channel through the yeast NPC (Shulga et al. 2000). Thus, in addition to having a structural role these NPC core components may also have a direct role in establishing the sieving properties of the NE and consequently in ensuring the correct segregation of nuclear and cytoplasmic processes.

3.1.3 Pore Membrane Proteins

In yeast there are only three integral membrane proteins found at the pore membrane (also commonly referred to as pore membrane proteins, or POMs): Pom152p, Pom34p, and Ndc1p (Wozniak et al. 1994; Chial et al. 1998; Rout et al. 2000). These proteins are characterized by their localization to or near the central plane of the NPC and by their ability to span the transcisternal pore membrane. Because of their localization they are generally assumed to be responsible for anchoring the NPC to the membrane. They might also have a central role in NPC biogenesis. For example, Pom152p is known to be able to form a stable ring-shaped substructure of the NPC that could be a stable NPC structural precursor (Strambio-de-Castillia et al. 1995). A recent topological and genetic study on Pom152p provides insights on the functional organization of this protein and on the way it interacts with other NPC components (Tcheperegine et al. 1999). The authors demonstrate that this protein spans the membrane only once and that the carboxyl-terminal bulk of the protein is positioned inside the lumen of the NE, while the amino-terminal domain faces the NPC. Thus, the amino-terminus is probably responsible for interacting with nucleoporins that compose the core domain of the NPC while the carboxyl terminus may be involved in pore formation and the maintenance of NPC structure. Interestingly, in addition to being a nucleoporin Ndc1p is also known to be part of the yeast spindle pole body (SPB; Winey et al. 1993; Chial et al. 1998). Pom34p is a very abundant constituent of highly-enriched yeast NPC fractions and like Pom152p, but contrary to Ndc1p, is non-essential (Winzeler et al. 1999; Rout et al. 2000). This result is surprising due to the postulated role of POMs in NPC biogenesis and because of the small number of POMs isolated to date. One possible explanation could be that other important players in the anchoring of NPCs to the NE have not been identified yet. This hypothesis is consistent with the observation that, unlike other nucleoporins, there is no obvious homology between the known yeast and vertebrate pore membrane proteins.

3.2 Other Conserved Domains Among Yeast Nucleoporins

Yeast nucleoporins also exhibit other conserved structural elements. Heptad repeats characteristic of coiled-coil domains are found in numerous FG nucleoporins and non-FG nucleoporins and have been implicated in protein–protein interactions (Grandi et al. 1993, 1995a,b; Hurwitz and Blobel 1995; Schlaich et al. 1997). Octapeptide motifs, termed nucleoporin RNA binding motifs or NRMs, are found within Nup100p, Nup116p, and Nup145p (Fabre et al. 1994). Such motifs have also been found in particular classes of RNA binding factors, but their functional relevance at the NPC remains unknown. An evolutionarily conserved region of 60 amino-acids, called the Gle2p binding sequence or GLEBS, is shared between yeast Nup116p and vertebrate NUP98 and mediates stable complex formation with Gle2p, a symmetrically positioned non-FG nucleoporin involved in mRNA export (Bailer et al. 1998; Ho et al. 1998; Rout et al. 2000).

Notably, however, sequence comparisons failed to identify possible ATPases or GTPases among yeast nucleoporins. This observation strongly suggests that yeast NPCs lack motor proteins or other components that could otherwise directly utilize energy to move substrates along the central transporter (Rout et al. 2000). In addition, no yeast nucleoporins (other than Nup2p, which is not essential in yeast) display obvious Ran binding domains. This observation indicates that Ran may not need to be bound to the NPC in the yeast *S. cerevisiae*, which is particularly interesting in light of the fact that several Ran binding domains are found on proteins associated with vertebrate NPCs. Thus, the localization of Ran to the NPC in vertebrates might represent an elaboration of nucleocytoplasmic transport not necessary for a possibly more streamlined mechanism utilized by the smaller and simpler yeast cells.

4 The Yeast NPC as a Dynamic Structure

4.1 Alterations of NPC Structure During Nucleocytoplasmic Transport

NPCs from different cell types undergo rearrangements during the passage of large transport substrates. A well known example is represented by the extensive rearrangements of by the nuclear basket during export of Balbiani ring mRNP particles through the NPC of *Chironomus* (Kiseleva et al. 1996). The central transporter, too, may dilate upon the transit of material through it (Akey and Goldfarb 1989; Akey 1990; Akey and Radermacher 1993; Kiseleva et al. 1998). As yeast NPCs share many structural features with metazoan NPCs it is likely that they also undergo structural alterations, similar to those described for metazoans. For example, the various classes of images reconstructed for yeast NPCs by Akey and co-workers could represent various conformations assumed by the NPCs during translocation of macromolecules in and out of the nucleus (Yang et al. 1998).

4.2 NPC Biogenesis

NPCs in all cell types have to be inserted in the plane of the NE upon their de novo biogenesis. Both in yeast and in other eukaryotes, NPCs appear to be assembled steadily throughout interphase (Maul 1977; Winey et al. 1997). Unlike yeast however, vertebrates and other metazoans are characterized by an "open mitosis", in which they break down their NE (and NPCs) during mitosis. In these systems, the reassembly of NPCs into the NE at telophase has thus been extensively studied to gain insight into the mechanism of NPC biogenesis. Multiple steps in the assembly process could be distinguished and ordered on the basis of their differential sensitivity to a variety of known assembly inhibitors, opening the road to the biochemical dissection of these intermediates (Macaulay and Forbes 1996). Furthermore, some intermediates have been imaged using scanning electron microscopy, suggesting that NPC assembly occurs at the NE in a step-by-step fashion (Goldberg et al. 1997). The first intermediates to be visualized are depressions in the outer nuclear membrane called "dimples". "Dimples" become holes that perforate the NE (called "empty" pores), which in turn serve as seeds for the subsequent formation of pre-NPC structures called "star rings" and "thin rings". It appears that in yeast, NPCs form by an extremely rapid process without any obvious intermediates (Winey et al. 1997). This process may therefore involve the formation of a nuclear pore of the correct diameter closely followed by the insertion of prefabricated NPC subcomplexes in the NE, which then mature into a functional NPC. A model has been suggested for how this process might occur for the duplication of the yeast SPB (Adams and Kilmartin 1999). Here, the formation of a pore across the NE and the insertion of the new SPB appear virtually simultaneous, preventing the formation of intermediate holes that could transiently alter NE permeability. A similarity between the mechanisms of SPB and NPC insertion into the NE is made more feasible by the fact that (at least) one of the NPC pore membrane proteins, Ndc1p, is also present in SPBs and is involved in SPB duplication (Winey et al. 1993; Chial et al. 1998). Other non-nucleoporins may also be involved in the process of NPC biogenesis at interphase. For example, this role has been suggested for Sec13p, a protein normally found in COPII vesicles trafficking between ER and Golgi but that was also found to have a functional role at the yeast NPC (Siniossoglou et al. 1996, 2000). Sec13p might be involved in stabilizing the reflexed membranes of the nascent nuclear pores or in promoting membrane fusion during NPC assembly; this would explain its dual localization at the NPCs and on the ER.

In a recent effort to identify factors that regulate yeast NPC formation and dynamics, Wente and coworkers employed a fluorescence-based strategy to distinguish between wild type yeast cells from mutants with impaired incorporation of a Green Fluorescent Protein (GFP)-tagged nucleoporin into NPCs (Bucci and Wente 1998). Significantly, using this fluorescence-based screen they identified an allele of a known nucleoporin (Nup57p) that resulted in lowered NPC assembly not only of Nup49p but also of two other FG nucleoporins, Nup116p and Nsp1p. Equally important is the observation that other nucleoporins (Nic96p, Nup8p, Nup159p, Nup145p, and Pom152p) did not appear to require the carboxyl terminus of Nup57p for their recruitment into NPCs. A third observation was that the nuclear import capacity of cells harboring this carboxyl terminal truncation of Nup57p was markedly diminished. These results suggest that there appears to be a hierarchy of nucleoporin recruitment into the NPCs in yeast and that certain nucleoporins are required to be present before other can associate with the nascent NPC.

4.3 Movement of Yeast NPCs in the NE

Heterogeneity in NPC spatial distribution on the surface of the NE has been observed both in wild-type yeast and higher eukaryotes cells and appears to depend upon the cell cycle, a variety of growth conditions, and the state of transcriptional activation of specific domains of the genome (Franke and Sheer 1974, and references therein; Winey et al. 1997). In addition, recent in vivo studies that employed GFP-labeled nucleoporins in combination with mating assays, have confirmed that yeast NPCs are highly dynamic structures capable of freely moving around in the plane of the NE (Belgareh and Doye 1997; Bucci and Wente 1997).

4.4 Regulation of Nucleocytoplasmic Transport During the Cell Cycle

As discussed above, some nuclear transport factors bind to preferred docking sites at the NPC. In the case of the transport factor Kap121p docking to Nup53p, this may also be a cell cycle-regulated interaction. Nup53p is phosphorylated in a cell cycle-specific manner, which coincides with a transient reduction of Kap121p binding to the NPC in vivo (Marelli et al. 1998). This might provide a means to regulate the import of Kap121p substrates into the nucleus during mitosis. Interestingly, a possible link between the cell cycle and the regulation of nucleocytoplasmic transport was also observed in metazoans (Feldherr and Akin 1991, 1993, 1994b) underscoring the universality of this phenomenon.

5 Putting It All Together: The Internal Organization of the Yeast NPC

5.1 Architectural Organization of Nucleoporins

A rough architectural map of the yeast NPC can be generated by combining data on biochemical and genetic interactions among nucleoporins with data on their localization to particular substructures of the NPC (Fig. 2; Kraemer et al. 1995; Nehrbass et al. 1996; Fahrenkrog et al. 1998; Hurwitz et al. 1998; Marelli et al. 1998; Kosova et al. 1999; Strahm et al. 1999; Bailer et al. 2000; Rout et al. 2000). This low resolution structural map could serve as a blueprint for drawing a detailed map of interactions among nucleoporins and between nucleoporins and their functional environment including transport factors and other cytoplasmic and nuclear structures. Consequently, the understanding of the architectural organization of the NPC could be used to derive conclusions on possible mechanisms of nucleocytoplasmic transport and of NPC biogenesis and dynamics.

One major conclusion that can be drawn from the yeast NPC structural map is that the symmetrical distribution of nucleoporins directly reflects the high

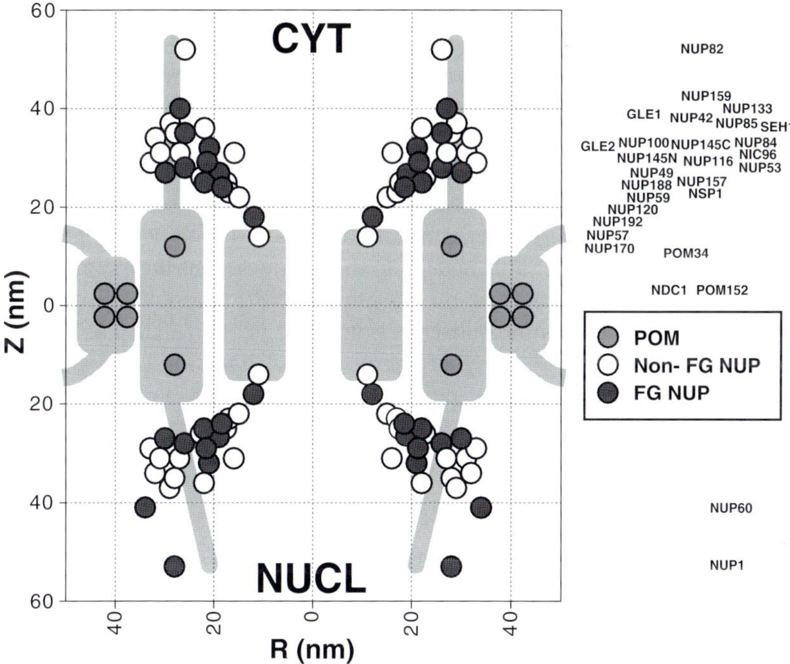

Fig. 2. Plot of the positions of nucleoporins in the yeast NPC. (Adapted from Rout et al. 2000)

level of symmetry displayed by the structure as a whole. Yeast nucleoporins belong to four distinct surface-accessible localization patterns: symmetrical, strictly cytoplasmic, strictly nuclear, and biased asymmetrical localization either to the cytoplasmic or to the nuclear sides of the NPC. The existence of nucleoporins localized exclusively to either the cytoplasmic fibrils or the nuclear basket was well established in a variety of model systems. Strikingly though, the results of the most recent analyses clearly document that such unequal distribution between the two sides of the NPC is an exception rather than a rule and is true only for a minority of yeast nucleoporins. Roughly one-third to one-fourth of the yeast NPC mass is composed of members of the FG family. The majority of these proteins are found on both sides of the NPC at roughly the same distance from the central plane of the NE (for example Nup49p, Nup57p, and Nsp1p). Thus, yeast NPCs are crowded with an abundance of FG docking sites positioned throughout the transport pathway across the NPCs, so that transport factors with their cognate substrate have the opportunity to interact with FG repeats at all stages of the translocation process. On the other hand, most of the nucleoporins that do display some level of asymmetry along the cylindrical axis of the NPC are also members of the FG family (such as for example the exclusively nucleoplasmic Nup60p; or Nup116p, which though on both sides of the NPC, is more abundant on the cytoplasmic face).

Hence, it seems safe to assume that this limited and organized asymmetry of transport factor docking sites is also important for the mechanism of transport. In particular it is worth noting that all three exclusively cytoplasmic nucleoporins (Nup159p, Nup82p, and Nup42p) have been functionally implicated in mRNA export (Kraemer et al. 1995; Neville et al. 1997; Stutz et al. 1997; Hurwitz et al. 1998; Rout et al. 2000); while at least one of the nucleoplasmic nucleoporins (Nup1p) has been found to bind preferentially to members of the β-karyopherin family, known to be involved in the import of proteins to the nucleus. This has led to the proposal that the asymmetrical yeast FG nucleoporins positioned on peripheral NPC filaments may be involved in the release of transport substrates en route to their cytoplasmic or nuclear destination from the vicinity of the pore (Belanger et al. 1994; Schlaich and Hurt 1995; Aitchison et al. 1996; Delphin et al. 1997; Rout et al. 1997, 2000; Hurwitz et al. 1998; Floer and Blobel 1999; Kehlenbach et al. 1999; Titov and Blobel 1999).

5.2 Interactions Between Nucleoporins

The discovery of distinct structural elements within the NPC architecture suggest the existence of multiple interconnecting NPC subcomplexes in which each specific nucleoporin form multiple connections with one another. To date, only four well-defined discrete NPC subcomplexes have been isolated from yeast.

The first of such subcomplexes to be identified was the Nsp1p–Nup57p–Nup49p–Nic96p complex generally thought to be the functional equivalent of the vertebrate p62 complex (Grandi et al. 1993, 1995b, 1997; Guan et al. 1995; Hu et al. 1996; Schlaich et al. 1997). With the exception of Nic96p, the members of this complex have a bipartite structure characterized by a FG domain at or near the amino-terminus, which interacts with various known transport factors and a coiled-coil domain at the carboxyl-terminus that is responsible for the formation of the complex. All the members of this complex are found on both sides of the NPC, close to the central plane of the NPC (Nehrbass et al. 1990; Rout et al. 2000). Interestingly, this observation is consistent with localization studies performed in vertebrates (Guan et al. 1995; Hu et al. 1996). In addition, in vivo functional data obtained with mutant forms of these proteins suggest a direct involvement of Nup49p and Nsp1p in transport across the NPC (Mutvei et al. 1992; Nehrbass et al. 1993; Doye et al. 1994). Taken together, these results argue that the Nsp1p complex represents a highly conserved subcomplex of the NPC that is most likely involved in the translocation of transport substrates across the central transporter.

Interestingly, four nucleoporins (Nup84p, Nup85p, Nup120p, and the carboxyl-terminal domain of Nup145p) have been found within a stable biochemical complex, the so called Nup84p complex, which also contains Sec13p and its homologue the nucleoporin Seh1p (Siniossoglou et al. 1996,; 2000).

Yeast strains harboring mutant versions of these four nucleoporins exhibit various NE aberrations such as clustered NPCs and NE blebs over the NPCs in addition to other defects connected to nuclear and nucleolar organization (for reviews see Doye and Hurt 1995; Wente et al. 1998). Strikingly, all of the nucleoporin mutant strains that are characterized by NPC-clustering also exhibit in vivo mRNA nuclear accumulation suggesting a possible role in nuclear export of mRNA (Doye et al. 1994; Fabre et al. 1994; Wente and Blobel 1994; Aitchison et al. 1995a; Gorsch et al. 1995; Li et al. 1995; Pemberton et al. 1995; Del Priore et al. 1996; Goldstein et al. 1996; Murphy et al. 1996). Two other nucleoporins exhibiting the twofold mRNA export/NPC clustering phenotype, Nup133p and Nup159p, are also known to genetically interact with a component of the Nup84p complex, namely Nup120p (Heath et al. 1995). The Nup84p complex was recently purified, and shown to exhibit a rough Y-shaped morphology and a molecular mass of ~375 kDa (Siniossoglou et al. 2000). This is significant because the Nup84p complex might represent an example of an NPC "building block", or stable structural module, with a distinct symmetrical localization and a function in both mRNA export and NE/NPC stability.

Interestingly, Nup82p and Nup159p form a complex with a pool of Nsp1p distinct from the one found in the Nsp1p–Nup57p–Nup49p–Nic96p complex. (Grandi et al. 1995a; Belgareh et al. 1998). This Nup82p–Nup159p–Nsp1p subcomplex may also interact with Nup116p which in turn interacts with Gle2p (Murphy et al. 1996; Bailer et al. 1998; Bailer et al. 2000). Mapping studies have demonstrated that Nsp1p and Nup159p bind to each other via their carboxyl-terminal coiled–coil regions and form the core of this complex. Nup82p interacts only with this preformed core complex but not with each protein individually. In addition, Nup116p appears to bind with the Nup82p complex via its carboxyl-terminal NRM domain.

The last stable yeast NPC subcomplex to be identified is the Nup53p complex, which is composed of the core component Nup170p and two novel FG nucleoporins, Nup53p and Nup59p (Marelli et al. 1998). The complex has a symmetrical distribution with respect to the mid plane of the NPC as demonstrated by IEM studies (Marelli et al. 1998; Rout et al. 2000) and was directly implicated in the process of active nucleocytoplasmic transport by virtue of its specific interaction with Kap121p, a member of the β-karyopherin family involved in the nuclear import of ribosomal proteins and of the transcription factor Pho4p (Rout et al. 1997; Kaffman et al. 1998). Interestingly, although each of the nucleoporins interacts with one another, the association of the Nup53p complex to Kap121p is specifically mediated by Nup53p alone. Moreover, Kap121p is the only β-type karyopherin known to specifically interact with Nup53p, suggesting that Nup53p is the primary docking site for this import factor. In addition, deletion of Nup53p from the yeast cell not only alters the subcellular localization of Kap121p but it also abrogates the Kap121p-mediated nuclear import of a reporter carrying the NLS for Rpl25p (the large subunit ribosomal protein, L25), underscoring the functional relevance of the interaction between Nup53p and this transport factor.

Several other interactions between individual nucleoporins have been reported. However, these interactions have not been completely characterized at the biochemical level and have not been shown to be necessarily direct. Examples of such interactions are represented by the Nup192–Nic96p complex (Kosova et al. 1999) and the interaction between Gle1p and Nup42p (Strahm et al. 1999).

Despite major progress in this field, several questions remain open. In particular, the way in which each individual subcomplex identified so far is integrated within the mass of the NPC and the network of interactions that hold each complex together are still poorly understood. Indeed, most nucleoporins have not yet been assigned to any subcomplex, and it is not known how the mirror- and bilaterally symmetrical eight spokes are held together (though this may involve homo-oligomeric interactions between adjacent nucleoporin subcomplexes). Consequently, a major effort needs to be carried out over the next several years in order to complete this task and to construct a fine structural map of yeast NPCs.

6 Functional Environment of the Yeast NPC: The Interactions with Its Neighbors

To properly function in the context of the living cell, the NPC cannot exist in isolation but it needs to functionally interact with its neighbors on both the cytoplasmic and the nucleoplasmic sides of the NE. Although very little is still known about the interactions between the NPC and cytoplasmic structures, some interesting connections have recently been unveiled that suggest that other surprises may be in store. For example, members of the Nup84p complex interact with Sec13p (see above) and with two ER membrane proteins involved in sporulation, Spo7p and Nem1p (Siniossoglou et al. 1998). The significance of these observations is not clear but it is possible that these proteins might take part with the NPC in the morphogenesis and structural maintenance of the NE.

On the nucleoplasmic side, NPCs are thought to be structurally and functionally continuous with the nuclear interior. Interconnecting open channels have been observed radiating from the nuclear interior towards NPCs (Berezney et al. 1995; Ris 1997), and the movement of proteins and RNAs along distinct intranuclear pathways can be studied both in vivo and in vitro (Lawrence et al. 1989; Xing et al. 1993; Zachar et al. 1993). Indeed, one model proposes the existence of a nucleoskeleton (defined as a non-chromatin intranuclear structural framework) composed of a filamentous network organized from the nuclear periphery and playing a major role in the direction of nuclear trafficking to and from the NPCs and in the functional organization of the chromatin in the nuclear interior (Berezney et al. 1995; Paddy 1998). Although the existence of tracks remains unproven, the investigation of the molecular basis of the connections between the NPCs and the nuclear interior has received great impetus from the study of the Tpr family of proteins (Kolling et al. 1993; Byrd et al. 1994; Bangs et al. 1996; Zimowska et al. 1997; Strambio-de-Castillia et al. 1999). These proteins are filamentous in nature (presumably because of the extensive domains of heptad repeats occupying the amino-terminal ~70% of the polypeptides) and extend from the tip of the NPC basket towards the nucleoplasm to form extensive networks of interconnecting filaments (Cordes et al. 1997; Strambio-de-Castillia et al. 1999). Both in metazoans and in yeast they have been implicated in the process of nucleocytoplasmic transport on the basis of functional studies (Bangs et al. 1998; Shah et al. 1998; Strambio-de-Castillia et al. 1999; Kosova et al. 2000). In addition, the Tpr NPC-association domain has been mapped to an amino-terminal domain of the protein in both yeast and vertebrates and a physical interaction between the yeast Tpr homologue, Mlp2p, and the nucleoporin, Nic96p has been observed (Kosova et al. 2000). Interestingly, new results point to a possible role of yeast Mlp1p and Mlp2p in regulating the functional architecture of chromatin inside the nucleus (Galy et al. 2000). In all cell types, genomic domains localized at or close to the NE are transcriptionally inactive or "silent". In yeast, these perinuclear chromatin regions are defined by the presence of telomeric clusters (Gottschling et al. 1990; Maillet et al. 1996; Andrulis et al. 1998). The only protein complex known to be involved in telomeric localization near the NE is yKu (the heterodimer, Yku70p/Yku80p), which has been implicated both in telomere maintenance and localization, and in DNA double-strand break repair (Boulton and Jackson 1996; Laroche et al. 1998; Martin et al. 1999; Mishra and Shore 1999; Milne et al. 1996). Yeast mutants lacking Mlp2p display a similar level of DNA double-strand break deficiency as yKu mutants. This phenotypic similarity is consistent with an apparent mislocalization of yKu in an Mlp2p mutant, and with the observation of an interaction between these two proteins. Strikingly, mutants null for both *MLP1* and *MLP2* also showed telomeric mislocalization and reduced telomeric silencing, similar to the null mutants of yKu. Finally, a genetic interaction between *MLP1* and the nucleoporin *NUP145* was also observed, suggesting that the role of Mlp1p and Mlp2p might be to define the perinuclear organization of silent telomeric DNA by tethering it to the NPCs via yKu (Galy et al. 2000). This model for the function of the yeast Tpr homologues does not exclude a concomitant role in nucleocytoplasmic transport and intranuclear trafficking. For example, by positioning silent chromatin close to regions of the NE near the NPCs, Mlp1p and Mlp2p might at the same time keep chromatin clear from the immediate vicinities of the nuclear baskets. This could in turn facilitate diffusion of material en route to the nuclear interior away from the pore, and also facilitate the NPC approach of export substrates (Strambio-de-Castillia et al. 1999).

7 Mechanism of Nucleocytoplasmic Transport

If the list of yeast NPC components is fairly comprehensive, and if our first views of its overall structure fairly accurate, then it is reasonable to speculate as to how the NPC functions as a macromolecular transporter. Several models have already been proposed (e.g., Koepp and Silver 1996). We can now limit these to a subset of possible models, by accepting that transport is not a motor-driven process (Rout et al. 2000), and that nucleotide hydrolysis is not required for translocation of many transport factor-cargo complexes across the NPC (Schwoebel et al. 1998; Englmeier et al. 1999). A key feature of the mechanism for transport appears to be the different affinities of the NPC binding sites for their cognate transport factor–cargo complexes. It is possible that Brownian motion moves each transport factor along a pathway of several preferred nucleoporin binding sites, from one face of the NE to the other (Blobel 1995); having thus traversed the NPC, exposure to the alternative Ran milieu from which the transport factor started (see above), terminates the transport reaction. The high number of different FG nucleoporins reflects the variety of transport factors, each following a different favored pathway of binding sites across the NPC. In this way, competition between the transport factors at the NPC is likely minimized (Aitchison et al. 1996; Rout et al. 1997; Marelli et al. 1998).

Three major principles may be involved in NPC-mediated transport. Firstly, an asymmetric distribution of binding sites between the nuclear and cytoplasmic faces of the NPC likely contributes (along with the Ran GTP/GDP gradient) to determining the directionality of transport for each transport factor, although the degree of involvement of such asymmetry for each transport factor is not clear (Gorlich et al. 1996; Delphin et al. 1997; Kehlenbach et al. 1999; Keminer et al. 1999; Nachury and Weis 1999). Secondly, dilation of the central transporter seems to occur during the passage of cargo across the NPC, but the degree to which this contributes to the gating function of the NPC for each transport pathway is unknown (Akey 1990; Akey and Radermacher 1993; Kiseleva et al. 1998; Yang et al. 1998). Thirdly, the confined central transporter presents a significant diffusion barrier to macromolecules. However, a large number of FG nucleoporins surround the transport path; thus, transport factor–cargo complexes that specifically dock to these nucleoporins can offset this diffusional exclusion with the energetically favorable process of binding, selecting for their efficient passage across the NPC. This principle has been used as the basis for a "virtual gating" model, but again, the degree of contribution of this process to actual NPC-mediated transport has yet to be established (Rout et al. 2000). These models are clearly simplified, and in particular it is likely that various transport factors differ in the detailed transport mechanisms they use. Determining the method each transport factor uses to traverse the NPC, and how this impacts the biology of the organism, remain major challenges for the next few years.

Acknowledgments. The authors thank John Aitchison for critical reading of the manuscript and many helpful suggestions, and Beth Hatton for excellent secretarial assistance. We gratefully acknowledge funding support from the Rockefeller University (C.S. and M.R.), and the Irma Hirschl Trust (M.R.).

References

Adams IR, Kilmartin JV (1999) Localization of core spindle pole body (SPB) components during SPB duplication in *Saccharomyces cerevisiae*. J Cell Biol 145:809–823

Aitchison JD, Blobel G, Rout MP (1995a) Nup120p: a yeast nucleoporin required for NPC distribution and mRNA transport. J Cell Biol 131:1659–1675

Aitchison JD, Rout MP, Marelli M, Blobel G, Wozniak RW (1995b) Two novel related yeast nucleoporins Nup170p and Nup157p: complementation with the vertebrate homologue Nup155p and functional interactions with the yeast nuclear pore-membrane protein Pom152p. J Cell Biol 131:1133–1148

Aitchison JD, Blobel G, Rout MP (1996) Kap104p: a karyopherin involved in the nuclear transport of messenger RNA binding proteins. Science 274:624–627

Akey CW (1990) Visualization of transport-related configurations of the nuclear pore transporter. Biophys J 58:341–355

Akey CW, Goldfarb DS (1989) Protein import through the nuclear pore complex is a multistep process. J Cell Biol 109:971–982

Akey CW, Radermacher M (1993) Architecture of the *Xenopus* nuclear pore complex revealed by three-dimensional cryo-electron microscopy. J Cell Biol 122:1–19

Andrulis ED, Neiman AM, Zappulla DC, Sternglanz R (1998) Perinuclear localization of chromatin facilitates transcriptional silencing [published erratum appears in Nature, 1 October 1998, 395(6701):525]. Nature 394:592–595

Bailer SM, Siniossoglou S, Podtelejnikov A, Hellwig A, Mann M, Hurt E (1998) Nup116p and nup100p are interchangeable through a conserved motif which constitutes a docking site for the mRNA transport factor gle2p. EMBO J 17:1107–1119

Bailer SM, Balduf C, Katahira J, Podtelejnikov A, Rollenhagen C, Mann M, Pante N, Hurt E (2000) Nup116p associates with the Nup82p–Nsp1p–Nup159p Nucleoporin Complex. J Biol Chem: 23540–23548

Bangs PL, Sparks CA, Odgren PR, Fey EG (1996) Product of the oncogene-activating gene Tpr is a phosphorylated protein of the nuclear pore complex. J Cell Biochem 61:48–60

Bangs P, Burke B, Powers C, Craig R, Purohit A, Doxsey S (1998) Functional analysis of Tpr: identification of nuclear pore complex association and nuclear localization domains and a role in mRNA export. J Cell Biol 143:1801–1812

Bayliss R, Ribbeck K, Akin D, Kent HM, Feldherr CM, Gorlich D, Stewart M (1999) Interaction between NTF2 and xFxFG-containing nucleoporins is required to mediate nuclear import of RanGDP. J Mol Biol 293:579–593

Belanger KD, Kenna MA, Wei S, Davis LI (1994) Genetic and physical interactions between Srp1p and nuclear pore complex proteins Nup1p and Nup2p. J Cell Biol 126:619–630

Belgareh N, Doye V (1997) Dynamics of nuclear pore distribution in nucleoporin mutant yeast cells. J Cell Biol 136:747–749

Belgareh N, Snay-Hodge C, Pasteau F, Dagher S, Cole CN, Doye V (1998) Functional characterization of a Nup159p-containing nuclear pore subcomplex. Mol Biol Cell 9:3475–3492

Berezney R, Mortillaro MJ, Ma H, Wei X, Samarabandu J (1995) The nuclear matrix: a structural milieu for genomic function. Int Rev Cytol 162 A:1–65

Blobel G (1995) Unidirectional and bidirectional protein traffic across membranes. Cold Spring Harbor Symp Quant Biol 60:1–10

Boulton SJ, Jackson SP (1996) Identification of a *Saccharomyces cerevisiae* Ku80 homologue: roles in DNA double strand break rejoining and in telomeric maintenance. Nucleic Acids Res 24: 4639–4648

Bucci M, Wente SR (1997) In vivo dynamics of nuclear pore complexes in yeast. J Cell Biol 136: 1185–1199

Bucci M, Wente SR (1998) A novel fluorescence-based genetic strategy identifies mutants of *Saccharomyces cerevisiae* defective for nuclear pore complex assembly. Mol Biol Cell 9:2439–2461

Buss F, Kent H, Stewart M, Bailer SM, Hanover JA (1994) Role of different domains in the self-association of rat nucleoporin p62. J Cell Sci 107:631–638

Byrd DA, Sweet DJ, Pante N, Konstantinov KN, Guan T, Saphire AC, Mitchell PJ, Cooper CS, Aebi U, Gerace L (1994) Tpr, a large coiled coil protein whose amino terminus is involved in activation of oncogenic kinases, is localized to the cytoplasmic surface of the nuclear pore complex. J Cell Biol 127:1515–1526

Chial HJ, Rout MP, Giddings TH, Winey M (1998) *Saccharomyces cerevisiae* Ndc1p is a shared component of nuclear pore complexes and spindle pole bodies. J Cell Biol 143:1789–1800

Cole CN, Hammell CM (1998) Nucleocytoplasmic transport: driving and directing transport. Curr Biol 8:R368–372

Cordes VC, Reidenbach S, Rackwitz HR, Franke WW (1997) Identification of protein p270/Tpr as a constitutive component of the nuclear pore complex-attached intranuclear filaments. J Cell Biol 136:515–529

Del Priore V, Snay CA, Bahr A, Cole CN (1996) The product of the *Saccharomyces cerevisiae* RSS1 gene, identified as a high-copy suppressor of the rat7-1 temperature-sensitive allele of the RAT7/NUP159 nucleoporin, is required for efficient mRNA export. Mol Biol Cell 7: 1601–1621

Delphin C, Guan T, Melchior F, Gerace L (1997) RanGTP targets p97 to RanBP2, a filamentous protein localized at the cytoplasmic periphery of the nuclear pore complex. Mol Biol Cell 8:2379–2390

Doye V, Hurt EC (1995) Genetic approaches to nuclear pore structure and function. Trends Genet 11:235–241

Doye V, Hurt E (1997) From nucleoporins to nuclear pore complexes. Curr Opin Cell Biol 9:401–411

Doye V, Wepf R, Hurt EC (1994) A novel nuclear pore protein Nup133p with distinct roles in poly(A)+ RNA transport and nuclear pore distribution. EMBO J 13:6062–6075

Dworetzky SI, Feldherr CM (1988) Translocation of RNA-coated gold particles through the nuclear pores of oocytes. J Cell Biol 106:575–584

Englmeier L, Olivo JC, Mattaj IW (1999) Receptor-mediated substrate translocation through the nuclear pore complex without nucleotide triphosphate hydrolysis. Curr Biol 9:30–41

Fabre E, Hurt E (1997) Yeast genetics to dissect the nuclear pore complex and nucleocytoplasmic trafficking. Annu Rev Genet 31:277–313

Fabre E, Boelens WC, Wimmer C, Mattaj IW, Hurt EC (1994) Nup145p is required for nuclear export of mRNA and binds homopolymeric RNA in vitro via a novel conserved motif. Cell 78:275–289

Fabre E, Schlaich NL, Hurt EC (1995) Nucleocytoplasmic trafficking: what role for repeated motifs in nucleoporins? Cold Spring Harbor Symp Quant Biol 60:677–685

Fahrenkrog B, Hurt EC, Aebi U, Pante N (1998) Molecular architecture of the yeast nuclear pore complex: localization of Nsp1p subcomplexes. J Cell Biol 143:577–588

Feldherr CM, Akin D (1991) Signal-mediated nuclear transport in proliferating and growth-arrested BALB/c 3T3 cells. J Cell Biol 115:933–939

Feldherr CM, Akin D (1993) Regulation of nuclear transport in proliferating and quiescent cells. Exp Cell Res 205:179–186

Feldherr CM, Akin D (1994a) Role of nuclear trafficking in regulating cellular activity. Int Rev Cytol 151:183–228

Feldherr CM, Akin D (1994b) Variations in signal-mediated nuclear transport during the cell cycle in BALB/c 3T3 cells. Exp Cell Res 215:206–210

Feldherr CM, Akin D (1997) The location of the transport gate in the nuclear pore complex. J Cell Sci 110:3065–3070

Feldherr CM, Kallenbach E, Schultz N (1984) Movement of a karyophilic protein through the nuclear pores of oocytes. J Cell Biol 99:2216–2222

Feldherr C, Akin D, Moore MS (1998) The nuclear import factor p10 regulates the functional size of the nuclear pore complex during oogenesis. J Cell Sci 111:1889–1896

Floer M, Blobel G (1996) The nuclear transport factor karyopherin beta binds stoichiometrically to Ran-GTP and inhibits the Ran GTPase activating protein. J Biol Chem 271:5313–5316

Floer M, Blobel G (1999) Putative reaction intermediates in Crm1-mediated nuclear protein export. J Biol Chem 274:16279–16286

Floer M, Blobel G, Rexach M (1997) Disassembly of RanGTP–karyopherin beta complex, an intermediate in nuclear protein import. J Biol Chem 272:19538–19546

Franke WW, Scheer U (1974) Structures and functions of the nuclear envelope. In: Busch H (ed) The cell nucleus. Academic Press, New York, pp 219–347

Galy V, Olivo-Marin JC, Scherthan H, Doye V, Rascalou N, Nehrbass U (2000) Nuclear pore complexes in the organization of silent telomeric chromatin (see comments). Nature 403:108–112

Goldberg MW, Allen TD (1992) High resolution scanning electron microscopy of the nuclear envelope: demonstration of a new, regular, fibrous lattice attached to the baskets of the nucleoplasmic face of the nuclear pores. J Cell Biol 119:1429–1440

Goldberg MW, Allen TD (1996) The nuclear pore complex and lamina: three-dimensional structures and interactions determined by field emission in-lens scanning electron microscopy. J Mol Biol 257:848–865

Goldberg MW, Wiese C, Allen TD, Wilson KL (1997) Dimples, pores, star-rings, and thin rings on growing nuclear envelopes: evidence for structural intermediates in nuclear pore complex assembly. J Cell Sci 110:409–420

Goldstein AL, Snay CA, Heath CV, Cole CN (1996) Pleiotropic nuclear defects associated with a conditional allele of the novel nucleoporin Rat9p/Nup85p. Mol Biol Cell 7:917–934

Gorlich D, Kutay U (1999) Transport between the cell nucleus and the cytoplasm. Annu Rev Cell Dev Biol 15:607–660

Gorlich D, Pante N, Kutay U, Aebi U, Bischoff FR (1996) Identification of different roles for RanGDP and RanGTP in nuclear protein import. EMBO J 15:5584–5594

Gorsch LC, Dockendorff TC, Cole CN (1995) A conditional allele of the novel repeat-containing yeast nucleoporin RAT7/NUP159 causes both rapid cessation of mRNA export and reversible clustering of nuclear pore complexes. J Cell Biol 129:939–955

Gottschling DE, Aparicio OM, Billington BL, Zakian VA (1990) Position effect at *S. cerevisiae* telomeres: reversible repression of Pol II transcription. Cell 63:751–762

Grandi P, Doye V, Hurt EC (1993) Purification of NSP1 reveals complex formation with "GLFG" nucleoporins and a novel nuclear pore protein NIC96. EMBO J 12:3061–3071

Grandi P, Emig S, Weise C, Hucho F, Pohl T, Hurt EC (1995a) A novel nuclear pore protein Nup82p which specifically binds to a fraction of Nsp1p. J Cell Biol 130:1263–1273

Grandi P, Schlaich N, Tekotte H, Hurt EC (1995b) Functional interaction of Nic96p with a core nucleoporin complex consisting of Nsp1p, Nup49p and a novel protein Nup57p. EMBO J 14:76–87

Grandi P, Dang T, Pane N, Shevchenko A, Mann M, Forbes D, Hurt E (1997) Nup93, a vertebrate homologue of yeast Nic96p, forms a complex with a novel 205-kDa protein and is required for correct nuclear pore assembly. Mol Biol Cell 8:2017–2038

Guan T, Muller S, Klier G, Pante N, Blevitt JM, Haner M, Paschal B, Aebi U, Gerace L (1995) Structural analysis of the p62 complex, an assembly of O-linked glycoproteins that localizes near the central gated channel of the nuclear pore complex. Mol Biol Cell 6:1591–1603

Heath CV, Copeland CS, Amberg DC, Del Priore V, Snyder M, Cole CN (1995) Nuclear pore complex clustering and nuclear accumulation of poly(A)+ RNA associated with mutation of the *Saccharomyces cerevisiae* RAT2/NUP120 gene. J Cell Biol 131:1677–1697

Hinshaw JE, Carragher BO, Milligan RA (1992) Architecture and design of the nuclear pore complex. Cell 69:1133–1141

Ho AK, Raczniak GA, Ives EB, Wente SR (1998) The integral membrane protein snl1p is genetically linked to yeast nuclear pore complex function. Mol Biol Cell 9:355–373

Hood JK, Silver PA (1998) Cse1p is required for export of Srp1p/Importin-alpha from the nucleus in *Saccharomyces cerevisiae* (in process citation). J Biol Chem 273:35142–35146

Hu T, Guan T, Gerace L (1996) Molecular and functional characterization of the p62 complex, an assembly of nuclear pore complex glycoproteins. J Cell Biol 134:589–601

Hurt E, Strasser K, Segref A, Bailer S, Schlaich N, Presutti C, Tollervey D, Jansen R (2000) Mex67p mediates nuclear export of a variety of RNA polymerase II transcripts. J Biol Chem 275:8361–8368

Hurwitz ME, Blobel G (1995) NUP82 is an essential yeast nucleoporin required for poly(A)+ RNA export. J Cell Biol 130:1275–1281

Hurwitz ME, Strambio-de-Castillia C, Blobel G (1998) Two yeast nuclear pore complex proteins involved in mRNA export form a cytoplasmically oriented subcomplex. Proc Natl Acad Sci USA 95:11241–11245

Iovine MK, Wente SR (1997) A nuclear export signal in Kap95p is required for both recycling the import factor and interaction with the nucleoporin GLFG repeat regions of Nup116p and Nup100p. J Cell Biol 137:797–811

Kaffman A, Rank NM, O'Shea EK (1998) Phosphorylation regulates association of the transcription factor Pho4 with its import receptor Pse1/Kap121. Genes Dev 12:2673–2683

Katahira J, Strasser K, Podtelejnikov A, Mann M, Jung JU, Hurt E (1999) The Mex67p-mediated nuclear mRNA export pathway is conserved from yeast to human. EMBO J 18:2593–2609

Kehlenbach RH, Dickmanns A, Kehlenbach A, Guan T, Gerace L (1999) A role for RanBP1 in the release of CRM1 from the nuclear pore complex in a terminal step of nuclear export (in process citation). J Cell Biol 145:645–657

Keminer O, Siebrasse JP, Zerf K, Peters R (1999) Optical recording of signal-mediated protein transport through single nuclear pore complexes. Proc Natl Acad Sci USA 96:11842–11847

Kenna MA, Petranka JG, Reilly JL, Davis LI (1996) Yeast Nle3p/Nup170p is required for normal stoichiometry of FG nucleoporins within the nuclear pore complex. Mol Cell Biol 16: 2025–2036

Kiseleva E, Goldberg MW, Daneholt B, Allen TD (1996) RNP export is mediated by structural reorganization of the nuclear pore basket. J Mol Biol 260:304–311

Kiseleva E, Goldberg M, Allen T, Akey C (1998) Active nuclear pore complexes in chironomous: visualization of transporter configurations related to mRNP export. J Cell Sci 111:223–236

Koepp DM, Silver PA (1996) A GTPase controlling nuclear trafficking: running the right way or walking randomly? Cell 87:1–4

Kolling R, Nguyen T, Chen EY, Botstein D (1993) A new yeast gene with a myosin-like heptad repeat structure. Mol Gen Genet 237:359–369

Kosova B, Pante N, Rollenhagen C, Hurt E (1999) Nup192p is a conserved nucleoporin with a preferential location at the inner site of the nuclear membrane. J Biol Chem 274:22646–22651

Kosova B, Pante N, Rollenhagen C, Podtelejnikov A, Mann M, Aebi U, Hurt E (2000) Mlp2p, a component of nuclear pore attached intranuclear filaments, associates with nic96p. J Biol Chem 275:343–350

Kraemer DM, Strambio-de-Castillia C, Blobel G, Rout MP (1995) The essential yeast nucleoporin NUP159 is located on the cytoplasmic side of the nuclear pore complex and serves in karyopherin-mediated binding of transport substrate. J Biol Chem 270:19017–19021

Laroche T, Martin SG, Gotta M, Gorham HC, Pryde FE, Louis EJ, Gasser SM (1998) Mutation of yeast Ku genes disrupts the subnuclear organization of telomeres. Curr Biol 8:653–656

Lawrence JB, Singer RH, Marselle LM (1989) Highly localized tracks of specific transcripts within interphase nuclei visualized by in situ hybridization. Cell 57:493–502

Li O, Heath CV, Amberg DC, Dockendorff TC, Copeland CS, Snyder M, Cole CN (1995) Mutation or deletion of the *Saccharomyces cerevisiae* RAT3/NUP133 gene causes temperature-dependent nuclear accumulation of poly(A)+ RNA and constitutive clustering of nuclear pore complexes. Mol Biol Cell 6:401–417

Macaulay C, Forbes DJ (1996) Assembly of the nuclear pore: biochemically distinct steps revealed with NEM, GTP gamma S, and BAPTA. J Cell Biol 132:5–20

Maillet L, Boscheron C, Gotta M, Marcand S, Gilson E, Gasser SM (1996) Evidence for silencing compartments within the yeast nucleus: a role for telomere proximity and Sir protein concentration in silencer-mediated repression. Genes Dev 10:1796–1811

Marelli M, Aitchison JD, Wozniak RW (1998) Specific binding of the karyopherin Kap121p to a subunit of the nuclear pore complex containing Nup53p, Nup59p, and Nup170p. J Cell Biol 143:1813–1830

Martin SG, Laroche T, Suka N, Grunstein M, Gasser SM (1999) Relocalization of telomeric Ku and SIR proteins in response to DNA strand breaks in yeast. Cell 97:621–633

Mattaj IW, Englmeier L (1998) Nucleocytoplasmic transport: the soluble phase. Annu Rev Biochem 67:265–306

Maul GG (1977) The nuclear and cytoplasmic pore complex: structure, dynamics, distribution and evolution. Int Rev Cytol Suppl 5:75–186

Milne GT, Jin S, Shannon KB, Weaver DT (1996) Mutations in two Ku homologs define a DNA end-joining repair pathway in *Saccharomyces cerevisiae*. Mol Cell Biol 16:4189–4198

Mishra K, Shore D (1999) Yeast Ku protein plays a direct role in telomeric silencing and counteracts inhibition by rif proteins. Curr Biol 9:1123–1126

Murphy R, Watkins JL, Wente SR (1996) GLE2, a *Saccharomyces cerevisiae* homologue of the *Schizosaccharomyces pombe* export factor RAE1, is required for nuclear pore complex structure and function. Mol Biol Cell 7:1921–1937

Mutvei A, Dihlmann S, Herth W, Hurt EC (1992) NSP1 depletion in yeast affects nuclear pore formation and nuclear accumulation. Eur J Cell Biol 59:280–295

Nachury MV, Weis K (1999) The direction of transport through the nuclear pore can be inverted. Proc Natl Acad Sci USA 96:9622–9627

Nehrbass U, Blobel G (1996) Role of the nuclear transport factor p10 in nuclear import. Science 272:120–122

Nehrbass U, Kern H, Mutvei A, Horstmann H, Marshallsay B, Hurt EC (1990) NSP1: a yeast nuclear envelope protein localized at the nuclear pores exerts its essential function by its carboxy-terminal domain. Cell 61:979–989

Nehrbass U, Fabre E, Dihlmann S, Herth W, Hurt EC (1993) Analysis of nucleo-cytoplasmic transport in a thermosensitive mutant of nuclear pore protein NSP1. Eur J Cell Biol 62:1–12

Nehrbass U, Rout MP, Maguire S, Blobel G, Wozniak RW (1996) The yeast nucleoporin Nup188p interacts genetically and physically with the core structures of the nuclear pore complex. J Cell Biol 133:1153–1162

Neville M, Stutz F, Lee L, Davis LI, Rosbash M (1997) The importin b family member Crm1p bridges the interaction between Rev and the nuclear pore complex during nuclear export. Curr Biol 7:767–775

Nigg EA (1997) Nucleocytoplasmic transport: signals, mechanisms and regulation. Nature 386: 779–787

Ohno M, Fornerod M, Mattaj IW (1998) Nucleocytoplasmic transport: the last 200 nanometers. Cell 92:327–336

Paddy MR (1998) The Tpr protein: linking structure and function in the nuclear interior? Am J Hum Genet 63:305–310

Paine PL, Moore LC, Horowitz SB (1975) Nuclear envelope permeability. Nature 254:109–114

Pemberton LF, Rout MP, Blobel G (1995) Disruption of the nucleoporin gene NUP133 results in clustering of nuclear pore complexes. Proc Natl Acad Sci USA 92:1187–1191

Pemberton LF, Rosenblum JS, Blobel G (1997) A distinct and parallel pathway for the nuclear import of an mRNA-binding protein. J Cell Biol 139:1645–1653

Pemberton LF, Blobel G, Rosenblum JS (1998) Transport routes through the nuclear pore complex. Curr Opin Cell Biol 10:392–399

Radu A, Blobel G, Moore MS (1995) Identification of a protein complex that is required for nuclear protein import and mediates docking of import substrate to distinct nucleoporins. Proc Natl Acad Sci USA 92:1769–1773

Rexach M, Blobel G (1995) Protein import into nuclei: association and dissociation reactions involving transport substrate, transport factors, and nucleoporins. Cell 83:683–692

Ris H (1991) The three dimensional structure of the nuclear pore complexes as seen by high voltage electron microscopy and high resolution low voltage scanning electron microscopy. EMSA Bull 21:54–56

Ris H (1997) High-resolution field-emission scanning electron microscopy of nuclear pore complex. Scanning 19:368–375

Rout MP, Blobel G (1993) Isolation of the nuclear pore complex. J Cell Biol 123:771–783

Rout MP, Wente SR (1994) Pores for thought: nuclear pore complex proteins. Trends Cell Biol 4: 357–365

Rout MP, Blobel G, Aitchison JD (1997) A distinct nuclear import pathway used by ribosomal proteins. Cell 89:715–725

Rout MP, Aitchison JD, Suprapto A, Hjertaas K, Zhao Y, Chait BT (2000) The yeast nuclear pore complex: composition, architecture, and transport mechanism. J Cell Biol 148:635–651

Ryan KJ, Wente SR (2000) The nuclear pore complex: a protein machine bridging the nucleus and cytoplasm [In process citation]. Curr Opin Cell Biol 12:361–371

Schlaich NL, Hurt EC (1995) Analysis of nucleocytoplasmic transport and nuclear envelope structure in yeast disrupted for the gene encoding the nuclear pore protein Nup1p. Eur J Cell Biol 67:8–14

Schlaich NL, Haner M, Lustig A, Aebi U, Hurt EC (1997) In vitro reconstitution of a heterotrimeric nucleoporin complex consisting of recombinant Nsp1p, Nup49p, and Nup57p. Mol Biol Cell 8:33–46

Schwoebel ED, Talcott B, Cushman I, Moore MS (1998) Ran-dependent signal-mediated nuclear import does not require GTP hydrolysis by Ran. J Biol Chem 273:35170–35175

Seedorf M, Damelin M, Kahana J, Taura T, Silver P (1999) Interactions between a nuclear transporter and a subset of nuclear pore complex proteins depend on Ran GTPase. Mol Cell Biol 19:1547–1557

Shah S, Tugendreich S, Forbes D (1998) Major binding sites for the nuclear import receptor are the internal nucleoporin Nup153 and the adjacent nuclear filament protein Tpr. J Cell Biol 141:31–49

Shulga N, Mosammaparast N, Wozniak R, Goldfarb DS (2000) Yeast nucleoporins involved in passive nuclear envelope permeability (in process citation). J Cell Biol 149:1027–1038

Siniossoglou S, Wimmer C, Rieger M, Doye V, Tekotte H, Weise C, Emig S, Segref A, Hurt EC (1996) A novel complex of nucleoporins, which includes Sec13p and a Sec13p homolog, is essential for normal nuclear pores. Cell 84:265–275

Siniossoglou S, Santos-Rosa H, Rappsilber J, Mann M, Hurt E (1998) A novel complex of membrane proteins required for formation of a spherical nucleus. EMBO J 17:6449–6464

Siniossoglou S, Lutzmann M, Santos-Rosa H, Leonard K, Mueller S, Aebi U, Hurt E (2000) Structure and assembly of the Nup84p complex. J Cell Biol 149:41–54

Solsbacher J, Maurer P, Bischoff FR, Schlenstedt G (1998) Cse1p is involved in export of yeast importin alpha from the nucleus. Mol Cell Biol 18:6805–6815

Strahm Y, Fahrenkrog B, Zenklusen D, Rychner E, Kantor J, Rosbach M, Stutz F (1999) The RNA export factor Gle1p is located on the cytoplasmic fibrils of the NPC and physically interacts with the FG-nucleoporin Rip1p, the DEAD-box protein Rat8p/Dbp5p and a new protein Ymr 255p. EMBO J 18:5761–5777

Strambio-de-Castillia C, Blobel G, Rout MP (1995) Isolation and characterization of nuclear envelopes from the yeast *Saccharomyces*. J Cell Biol 131:19–31

Strambio-de-Castillia C, Blobel G, Rout MP (1999) Proteins connecting the nuclear pore complex with the nuclear interior. J Cell Biol 144:839–855

Stutz F, Kantor J, Zhang D, McCarthy T, Neville M, Rosbash M (1997) The yeast nucleoporin rip1p contributes to multiple export pathways with no essential role for its FG-repeat region. Genes Dev 11:2857–2868

Tcheperegine SE, Marelli M, Wozniak RW (1999) Topology and functional domains of the yeast pore membrane protein Pom152p. J Biol Chem 274:5252–5258

Titov AA, Blobel G (1999) The karyopherin Kap122p/Pdr6p imports both subunits of the transcription factor IIA into the nucleus. J Cell Biol 147:235–246

Unwin PN, Milligan RA (1982) A large particle associated with the perimeter of the nuclear pore complex. J Cell Biol 93:63–75

Wente SR (2000) Gatekeepers of the nucleus. Science 288:1374–1377

Wente SR, Blobel G (1994) NUP145 encodes a novel yeast glycine–leucine–phenylalanine–glycine (GLFG) nucleoporin required for nuclear envelope structure. J Cell Biol 125:955–969

Wente SR, Gasser SM, Caplan AJ (1998) The nucleus and nucleocytoplasmic transport in *Saccharomyces cerevisiae*. In: Broach JR, Jones E, Pringle J (eds) The molecular and cellular biology of the yeast *Saccharomyces*. Cold Spring Harbor Press, Cold Spring Harbor, NY

Winey M, Hoyt MA, Chan C, Goetsch L, Botstein D (1993) NDC1: a nuclear periphery component required for yeast spindle pole body duplication. J Cell Biol 122:743–751

Winey M, Yarar D, Giddings TH Jr, Mastronarde DN (1997) Nuclear pore complex number and distribution throughout the *Saccharomyces cerevisiae* cell cycle by three-dimensional reconstruction from electron micrographs of nuclear envelopes. Mol Biol Cell 8:2119–2132

Winzeler EA, Shoemaker DD, Astromoff A, Liang H, Anderson K, Andre B, Bangham R, Benito R, Boeke JD, Bussey H, Chu AM, Connelly C, Davis K, Dietrich F, Dow SW, El Bakkoury M, Foury F, Friend SH, Gentalen E, Giaever G, Hegemann JH, Jones T, Laub M, Liao H, Davis RW et al. (1999) Functional characterization of the *S. cerevisiae* genome by gene deletion and parallel analysis. Science 285:901–906

Wozniak RW, Blobel G, Rout MP (1994) POM152 is an integral membrane protein of the pore membrane domain of the nuclear envelope. J Cell Biol 125:31–42

Wozniak RW, Rout MP, Aitchison JD (1998) Karyopherins and kissing cousins. Trends Cell Biol 8:184–188

Xing Y, Johnson CV, Dobner PR, Lawrence JB (1993) Higher level organization of individual gene transcription and RNA splicing. Science 259:1326–1330

Yang Q, Rout MP, Akey CW (1998) Three-dimensional architecture of the isolated yeast nuclear pore complex: functional and evolutionary implications. Mol Cell 1:223–234

Zabel U, Doye V, Tekotte H, Wepf R, Grandi P, Hurt EC (1996) Nic96p is required for nuclear pore formation and functionally interacts with a novel nucleoporin, Nup188p. J Cell Biol 133:1141–1152

Zachar Z, Kramer J, Mims IP, Bingham PM (1993) Evidence for channeled diffusion of pre-mRNAs during nuclear RNA transport in metazoans. J Cell Biol 121:729–742

Zimowska G, Aris JP, Paddy MR (1997) A *Drosophila* Tpr protein homolog is localized both in the extrachromosomal channel network and to nuclear pore complexes. J Cell Sci 110:927–944

The Vertebrate Nuclear Pore Complex: From Structure to Function

Birthe Fahrenkrog[1]* and Ueli Aebi[2]

1 Introduction

In interphase eukaryotic cells, the cytoplasm is spatially separated from the cell nucleus by the double-membraned nuclear envelope (NE). Inserted into the NE are nuclear pore complexes (NPCs) which mediate bidirectional nucleocytoplasmic transport, i.e. the passive diffusion of ions and small molecules as well as the signal-mediated transport of proteins, RNAs and RNP particles in and out of the nucleus (reviewed in Izaurralde and Adam 1998; Mattaj and Englmeier 1998; Ohno et al. 1998; Görlich and Kutay 1999). Extensive electron microscopy studies, predominantly on *Xenopus* oocytes, have revealed the three-dimensional (3-D) architecture of the NPC (reviewed in Panté and Aebi 1996a; Stoffler et al. 1999a; Allen et al. 2000; Fahrenkrog et al. 2001). Accordingly, the vertebrate NPC reveals an eight-fold symmetric (i.e. in the plane of the NE) tripartite (i.e. perpendicular to the plane of the NE) architecture with a total mass of ~125 MDa. The ~55 MDa central framework of the NPC (also called the spoke complex) is built of eight multi-domain spokes, each consisting of two roughly identical halves (Fig. 1). The central framework is sandwiched between a ~32 MDa cytoplasmic and a ~21 MDa nuclear ring moiety. Eight short kinky fibrils emanate from the cytoplasmic ring, protruding into the cytoplasm, whereas the nuclear ring is capped by a basket-like assembly (also called the fish-trap), made of eight thin fibrils that join distally so as to form an iris-like ring.

The ring-like, quasi-822 symmetric central framework of the NPC harbors a central pore, 40–50 nm in diameter, which represents the exclusive gateway for signal-mediated nucleocytoplasmic transport. By electron microscopy this central pore appears frequently plugged by a particle of variable size and shape. The central plug, rather than representing a stationary component of the NPC, may in fact be a mobile moiety, for example, cargo caught in transit (Reichelt et al. 1990; Jarnik and Aebi 1991; Stoffler et al. 1999a). This overall 3-D architecture of the NPC appears to be shared among distinct species,

* *Present address:* M.E. Müller Institute for Structural Biology, Biozentrum, University of Basel, Klingelbergstraße 70, 4056 Basel, Switzerland

[1] EMBL, Meyerhofstrasse 1, 69117 Heidelberg, Germany

[2] M.E. Müller Institute for Structural Biology, Biozentrum, University of Basel, Klingelbergstraße 70, 4056 Basel, Switzerland

Results and Problems in Cell Differentiation, Vol. 35
K. Weis (Ed.): Nuclear Transport

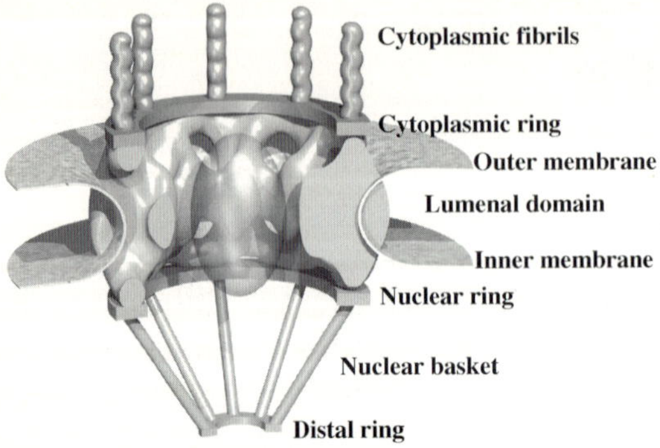

Fig. 1. Schematic representation of the 3-D architecture of the NPC with distinct components indicated

although, for example, the yeast NPC appears to be ~15% smaller in its linear dimensions (Fahrenkrog et al. 1998) with a molecular mass of only about 60 MDa (Rout and Blobel 1993; Yang et al. 1998; Rout et al. 2000).

2 Toward the Three-Dimensional Architecture of the Nuclear Pore Complex

For a structure-based functional understanding of nucleocytoplasmic transport, determination of the 3-D architecture of the NPC and of its molecular composition are mandatory. The 3-D architecture of the central framework of the NPC has been determined by 3-D reconstructions of both negatively stained and frozen-hydrated *Xenopus* oocyte NEs after detergent treatment (Hinshaw et al. 1992; Akey and Radermacher 1993). Both 3-D reconstructions revealed the central framework to consist of eight multi-domain spokes built of two approximately identical halves (see Fig. 1; adapted from Hinshaw et al. 1992). Each spoke is built of two copies each of four distinct morphological domains, termed the annular, column, lumenal and ring domains (see Hinshaw et al. 1992), from which the lumenal domain protrudes into the lumen of the NE. The 3-D reconstruction determined from negatively stained NPCs unveiled eight distinct peripheral channels of ~10 nm diameter each (see Fig. 1; Hinshaw et al. 1992), which were less obvious – if not absent – in the 3-D reconstruction obtained from ice-embedded NPCs (Akey and Radermacher 1993). It has been speculated that these distinct peripheral channels may represent gateways for the passive diffusion of small molecules and ions (see below). In contrast, signal-mediated nucleocytoplasmic transport of proteins, RNAs and RNP particles occurs via the ~45-nm-diameter central pore that lines the eightfold symmetry axis of the central framework.

Eight distinct peripheral channels were also delineated in a tomographic 3-D reconstruction of completely native (i.e. without any chemical fixation and

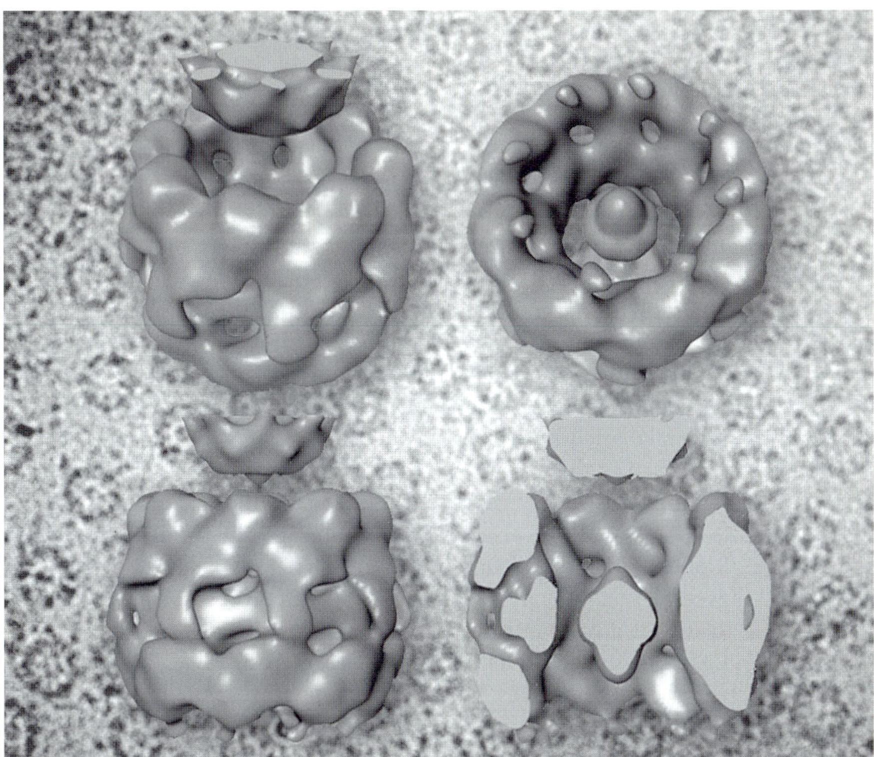

Fig. 2. Three-dimensional reconstruction of *Xenopus* NPCs by cryo-electron tomography superimposed onto an electron micrograph of thick-ice embedded, completely unfixed and unstained NEs. *Top left* Slightly tilted en face view of the 3-D map revealing peripheral holes, the distal ring floating above the basic framework, as well as remnants of the putative intranuclear filaments and the basket filaments. *Top right* View from the cytoplasmic face unveiling the remnants of the cytoplasmic fibrils and the central plug. *Bottom left* En face view of the 3-D map revealed the distal ring including the attachment sites of the intranuclear filaments and the basket fibrils, the nuclear ring, the peripheral holes, the cytoplasmic ring as well as the remnants of the cytoplasmic fibrils. *Bottom right* A cut-open view of the 3-D map revealing distinct handles, the central plug and the peripheral holes and channels

detergent treatment) *Xenopus* oocyte NPCs that were embedded in a thick (i.e. 250 nm) amorphous ice film (Fig. 2; Stoffler et al. 2001). Moreover, in this latter 3-D reconstruction, which represents an average over 500 NPCs, eight distinct "handles" extending from the central framework into the lumen of the NE were depicted (Fig. 2). Theses handles might represent anchor sites for the NPC within the NE.

The recently determined 3-D structure of isolated yeast NPCs (Yang et al. 1998) reveals the same pronounced 822 symmetry as does the central framework of vertebrate NPCs. From the relatively flat overall structure of the yeast NPC, it appears that it lacks most of the distinct cytoplasmic and nuclear ring moieties of vertebrate NPCs (Figs. 1 and 2; see below). Moreover, judged from its overall linear dimensions, which are typically 15% smaller that those of the vertebrate NPC (see Fahrenkrog et al. 1998; Stoffler et al. 1999a), the yeast NPC has a mass of only about 60 MDa, a value that has experimentally been confirmed (Yang et al. 1998).

In contrast to the quasi-822 symmetric central framework of the *Xenopus* oocyte NPC, its cytoplasmic and nuclear periphery appear rather distinct. As depicted schematically in Fig. 1, the cytoplasmic ring moiety of the vertebrate NPC is decorated by eight short wavy fibrils, whereas the nuclear ring moiety is capped by a basket-like assembly consisting of eight tenuous fibrils that join distally to form a relatively massive ring. For the first time, this distal ring was resolved in a tomographic 3-D reconstruction of fully native, thick-ice-embedded *Xenopus* oocyte NPCs (Stoffler at al. 2001). As documented in Fig. 2, it reveals eight anchor sites each for the struts of the nuclear basket, which are not resolved in this reconstruction, and for filaments projecting into the nuclear interior, most likely representing the recently described intranuclear filaments (see Cordes et al. 1997). Remnants of the eight cytoplasmic fibrils were resolved too in this tomographic 3-D reconstruction, but due to their high degree of mechanical flexibility (see Panté and Aebi 1996a) it was not possible to depict anything beyond their anchor sites at the cytoplasmic ring (Fig. 2). Last, but not least, as can be gathered from Fig. 2, the cytoplasmic and nuclear rings themselves, while appearing similar at first glance, are clearly distinct in terms of their overall size and shape.

The structural asymmetry of the two peripheries of the NPC might be related to its functional asymmetry, as both nuclear import and export involve vectorial cargo movement. In this context, the cytoplasmic fibrils may acts as initial docking sites for nuclear import cargoes, whereas the nuclear basket, and in particular its distal ring, may serve as initial docking site for nuclear export cargoes. Whereas the nuclear basket appears to anchor distinct intranuclear filaments, for example, consisting of Tpr as their major molecular constituent (see Cordes et al. 1997; and see below), it is conceivable that the cytoplasmic fibrils might connect the NPC to cytoskeletal elements (Davis 1995). Both the cytoplasmic fibrils and the nuclear basket are highly flexible structures. The cytoplasmic fibrils, charged with docked import cargo, can bend so as to deliver the cargo complex to higher-affinity binding sites residing near or at the cytoplasmic entry to the central pore (Panté and Aebi 1996b). Similarly, the cytoplasmic fibrils may bind cargo as it exits the central pore, thereby acting as a terminal docking site for the export cargo complex to assure its orderly dissociation and controlled release of the cargo into the cytoplasm. Even more so, the nuclear basket which, via its distal ring, can open and close (see Fig. 4 and below), is predestined to serve as an initial docking site for export cargo and as a terminal docking site for import cargo. Also, it is conceivable that the rather massive distal ring (see Fig. 2), facilitated by highly flexible struts, may move axially so as to plug/unplug the nuclear opening of the central pore.

3 Molecular Constituents of the Nuclear Pore Complex: The Nucleoporins

Based on the molecular mass of ~125 MDa and the 822 symmetry of the central framework of the vertebrate NPC, it is assumed that it is composed of 8–16 copies of 50–100 different proteins, called nucleoporins (Nups; reviewed in Panté and Aebi 1996a; Stoffler et al. 1999a; Fahrenkrog et al. 2001). As yet, it is still unknown, if as in the ribosome, RNAs contribute to the structural integrity of the NPC. Whereas completion of the yeast genome project in combination with proteomics has led to the identification and initial characterization of presumably all yeast nucleoporins (Rout et al. 2000; see also Rout et al., this Vol.), in the case of the vertebrate NPC less than 20 nucleoporins have been identified and molecularly characterized (Table 1). Somewhat surprisingly, immunolocalization of the yeast nucleoporins via genomic protein-A tagging of their N- or C-termini has revealed them predominantly as constituents of the cytoplasmic or the nuclear periphery of the NPC (Rout et al. 2000). As documented in Fig. 3, a similar picture has emerged from immunolocalization of about one dozen vertebrate nucleoporins by the use of nucleoporin-specific antibodies. Hence we are faced with the apparent dilemma that molecularly the central framework of the NPC, for both yeast and vertebrate species, represents to a large extent a "black hole". This rather biased picture will, no doubt, slowly but definitely change as more elaborate 3-D structural

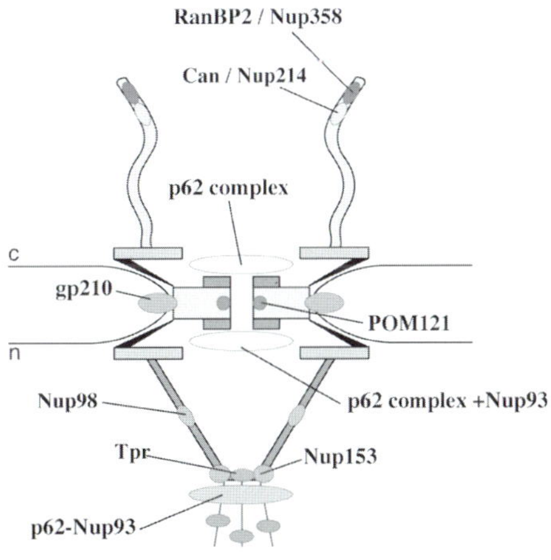

Fig. 3. Schematic diagram summarizing the immunolocalization of vertebrate nucleoporins by antibody-gold labeling of nucleoporin epitopes. Can/Nup214 and RanBP2/Nup358 exhibit epitopes at the cytoplasmic fibrils. The p62 complex resides at the cytoplasmic and nuclear periphery of the central channel. p62 is additionally located at the distal ring of the nuclear basket in complex with Nup93. Nup93 exhibits an additional epitope at the nuclear periphery of the central channel. Nup96, Nup98, Nup153 and Tpr are part of the nuclear basket, the latter also at the intranuclear filaments. The integral membrane proteins gp210 and POM121 are predicted to have epitopes in the lumen of the NE and in the NPC core

Table 1. Vertebrate nucleoporins

Name[a]	Homologue(s)	Motifs[b]	Location	Properties and function	References
p45	Sc Nup49p	FG, coiled coil	Cytoplasmic and nuclear periphery of the central channel	Generated by alternative splicing of p58; role in nuclear protein import	Panté and Aebi (1996a); Stoffler et al. (1999a); Hu and Gerace (1998)
p54	Sc Nup57p	FG, PA, coiled coil	Cytoplasmic and nuclear periphery of the central channel	Role in nuclear protein import	Panté and Aebi (1996a); Stoffler et al. (1999a); Hu and Gerace (1998)
p58	Sc Nup49p	FG, PA, coiled coil	Cytoplasmic and nuclear periphery of the central channel	Role in nuclear protein import	Panté and Aebi (1996a); Stoffler et al. (1999a); Hu and Gerace (1998)
p62	Sc Nsp1p, Hv p62	FXFG, coiled coil	Cytoplasmic and nuclear periphery of the central channel; nuclear basket	In complex with p45, p54 and p58; role in nuclear protein import	Panté and Aebi (1996a); Fischer et al. (1997)
Nup88	rNup84	Coiled coil	Cytoplasmic face of the NPC	C-terminal domain contains CAN/Nup214 binding site	Fornerod et al. (1997); Bastos et al. (1997)
Nup93	Sc Nic96p, Sp Npp106	Coiled coil	Nuclear periphery of the central channel; nuclear basket	Role in NPC assembly; in complex with p205	Grandi et al. (1997)
Nup96	Sc C-Nup145p	–	Nucleoplasmic face	Generated by autoproteolytic in vivo cleavage of a Nup98-Nup96 precursor; in complex with Nup107	Fontoura et al. (1999); Rosenblum and Blobel (1999)
Nup98	ScNup100p, Nup 116p, and N-Nup145p	FXFG, GLFG, FG	Nuclear basket and nucleus	Role in export of snRNAs, 5S RNA, rRNA, and mRNA,but not tRNA; role in import and export of HIV proteins; role in AMLs	Panté and Aebi (1996a); Powers et al. (1997); Kasper et al. (1999)
Nup107	Sc Nup84p	Leucine zipper	Cytoplasmic face		Panté and Aebi (1996a); Radu et al. (1994)
Pom121	–	FXFG, transmembrane	NPC core	Anchors NPC to the NE; N-terminal domain required for nuclear targeting, N-terminal and transmembrane domain required for NPC targeting	Panté and Aebi (1996a); Söderqvist et al. (1997)

Nup153	Sc Nup1p	FXFG, 4 Zn fingers	Nuclear basket	Termination site for nuclear protein import; N-terminus contains targeting and assembly information	Panté and Aebi (1996a); Shah et al. (1998); Enarson et al. (1998)
Nup155	D Nup154, Sc Nup 157p and Nup 170p	–	Cytoplasmic and nuclear face of the NPC		Panté and Aebi (1996); Gigliotti et al. (1998)
gp210	–	Transmembrane	Lumen of the NE	Anchors NPC to the NE; related to autoimmune diseases	Panté and Aebi (1996a)
CAN/ Nup214		FXFG, FG, leucine zipper	Cytoplasmic fibrils	Role in nuclear protein import, mRNA export and cell cycle; involved in AMLs	Panté and Aebi (1996a); Boer et al. (1997); Fornerod et al. (1997); Bastos et al. (1998)
Tpr (265 kDa)	Sc Mlp1p, Sc Mlp2p	Coiled coil	Nuclear basket and intranuclear filaments	C-terminus essential for nuclear import; N-terminus required for NPC association; possible role in mRNA export or recycling of transport factors; appears in oncogenic fusions with the oncogenes *met*, *trk*, and *raf*	Panté and Aebi (1996a); Cordes et al. (1997, 1998); Zimowska et al. (1997); Bangs et al. (1998); Kosova et al. (1999); Strambio-de-Castilla et al. (1999)
RanBP2/ Nup358	–	Ran binding, FXFG, FG, 8 Zn fingers	Cytoplasmic fibrils	Nucleocytoplasmic transport	Panté and Aebi (1996a)
PBC68	–	–	Nucleoplasmic face of NPC colocalizes to mitotic spindle	Involved in primary biliary cirrhosis	Theodoropoulos et al. (1999)

D, *Drosophila melanogaster*; h, human; Hv, *Hydra vulgaris*; Nup, nuclear pore protein, r, rat; Sc, *Saccharomyces cerevisiae*; Sp, *Schizosaccharomyces pombe*.

[a] Numerical assignment reflects either the predicted molecular mass (in kDa) or the genetic identification.

[b] FXFG, GLFG, FG, GSXS, GSSX, GFXS, PA, and WD are repeat motifs represented by single-letter code for amino acids; coiled coil, predicted parallel two-stranded α-helical structure made of heptad repeats.

analyses of yeast and vertebrate NPCs, and of distinct NPC subcomplexes are completed. Ultimately, crystal structures of individual nucleoporins and of nucleoporin complexes will emerge, so that these can be directly fitted into increasingly higher resolution, tomographic 3-D reconstructions of intact NPCs (see Fig. 2) and of distinct NPC subcomplexes. Last, but not least, with the completion of the mouse and human genome projects, identification and characterization of vertebrate nucleoporins should progress much more rapidly.

A common feature of a number of yeast and vertebrate nucleoporins are FG (Phe-Gly) repeat motifs within their amino acid sequence, a "signature" which might help in identifying new nucleoporins at the primary sequence level by genomic and/or proteomic approaches. Whereas these FG repeats are not required for NPC targeting to the NE and, in most cases, are not essential for viability, evidence has been accumulating that these sequence repeat motifs play a functional role in nucleocytoplasmic transport (see Rexach and Blobel 1995). Whereas blot overlays or solution binding assays appear to be rather unspecific for monitoring interactions between FG-containing nucleoporins and transport factors (see Radu et al. 1995), *Xenopus* egg extracts or isolated rat liver nuclei could demonstrate more specific interactions between distinct nucleoporins and transport factors (Shah et al. 1998; Shah and Forbes 1998). Most likely, FG repeat motifs may serve as "docking" sites for cargo complexes near or at the entry or exit site of the NPC's central pore. In addition, FG repeats may also function as "parking" sites to line up cargo at the cytoplasmic or nuclear NPC periphery for subsequent funneling to and passage through the central pore.

An in vivo interaction between Nup153 and the import receptor importin-β has been demonstrated in *Xenopus* egg extracts (Shah et al. 1998). Nup153 was first identified in rat (Sukegawa and Blobel 1993) and in humans (McMorrow et al. 1994). Since Nup153 exhibits epitopes near or at the distal ring of the nuclear basket (see Fig. 3; Panté et al. 1994, 2000), it is conceivable that Nup153 acts as termination site for nuclear protein import before the cargo complex is dissociated and the cargo released into the nucleus (Shah et al. 1998). However, Nup153 has also been demonstrated to participate in mRNA export (Bastos et al. 1996), most likely serving as the initial docking site for mRNA on its passage through the NPC.

Tpr, a constituent of the distal end of the nuclear basket and/or the intranuclear filaments associated with the nuclear basket via its distal ring (see Figs. 2 and 3; Cordes et al. 1997), also interacts with the import factors importin-α and importin-β (Bangs et al. 1998; Shah et al. 1998). This interaction only takes place in the absence of karyophilic proteins, suggesting that Tpr is involved in the recycling of importin-α and importin-β rather than in the actual import of karyophilic proteins into the nucleus(Bangs et al. 1998; Shah et al. 1998). Two yeast homologues of Tpr, Mlp1p and Mlp2p, have also been identified, and both proteins were located to the intranuclear filaments by immunogold-electron microscopy (Strambio-de-Castillia et al. 1999; Kosova et al.2000). Similar to Tpr, Mlp1p and Mlp2p are not involved in nuclear protein import, but whereas Tpr has also been suggested to participate in mRNA export (Bangs et al. 1998), this transport activity has not been evaluated for Mlp1p and Mlp2p (Strambio-de-Castilla et al. 1999).

The two nucleoporins Nup96 and Nup98 are generated by cleavage of a ~186-kDa precursor protein by an autocatalytic process without participation of any endogenous proteases (Fontoura et al. 1999; Rosenblum and Blobel 1999). The putative yeast homologue of Nup96/Nup98 is Nup145p, which is cleaved in vivo so as to yield two functionally distinct domains (Dockendorff et al. 1997; Emtage et al. 1997; Teixeira et al. 1997) termed C-Nup145p and N-Nup145p (for C-terminal domain of Nup145p and N-terminal domain of Nup145p). Nup98 appears to be the vertebrate homologue of yeast N-Nup145p, whereas Nup96 appears to be the vertebrate homologue of yeast C-Nup145p. While Nup98 and Nup96 have both been located to the nuclear basket (Fig. 3; see Radu et al. 1995; Fontoura et al. 1999), the yeast homologues have both been located simultaneously to the cytoplasmic and the nuclear periphery of the yeast NPC (Rout et al. 2000).

Nup98 is an O-linked glycoprotein that was first identified in *Xenopus* and rat (Powers et al. 1995; Radu et al. 1995). As a constituent of the nuclear basket, Nup98 appears to be primarily involved in distinct RNA export pathways, i.e. in the export of snRNAs, 5S RNA, rRNA and mRNA, but also in nuclear growth and replication (Powers et al. 1995, 1997; Radu et al. 1995). Nup98 is an FG-repeat containing nucleoporin, and its FG repeats have recently been reported to be involved in therapy-related and de novo forms of acute myeloid leukemia (AML) or myelodysplastic syndromes (MDS; Nakamura et al. 1996, 1999; Arai et al. 1997; Raza-Egilmez et al. 1998; Ahuja et al. 1999; Borrow et al. 1996; Ikeda et al. 1999; Kasper et al. 1999; Kwong and Pang, 1999; Nishiyama et al. 1999). Here the *NUP98* gene is mainly found in three distinct chromosomal translocations: the t(7;11)(p15;p15), the t(2;11)(q31;p15), and the inv(11)(p15;q22) translocation (Borrow et al. 1996; Nakamura et al. 1996; Arai et al. 1997; Raza-Egilmez et al.1998). Chimeric mRNAs produced by the t(7;11) and the t(2:11) translocation fuse the Nup98 N-terminal FG repeats with the homeodomains of HOXA9 andHOXD13, whereas the inv(11) translocation fuses the FG repeats of Nup98 with DDX10, a putative RNA helicase (Nakamura et al. 1996, 1999; Arai et al. 1997; Raza-Eglimez et al. 1998; Borrow et al. 1996; Ikeda et al. 1999; Kasper et al. 1999; Kwong and Pang, 1999; Nishiyama et al. 1999; Hatano et al. 2000). The mechanism by which these fusions become leukemogenic is not known, but it has been suggested that the resulting chimeric proteins consisting of the Nup98 FG repeats and the homeodomain of the homeobox proteins might function as oncogenic transcription factors due to the transactivation capacity of the *NUP98* FG repeats (Kasper et al. 1999; Nakamura et al. 1999). As an alternative, *NUP98* might act as a potential tumor suppressor gene (Borrow et al. 1996; Nakamura et al. 1999). The number of described chromosomal translocations including *NUP98* is still increasing (e.g., Ahuja et al. 1999; Nakamura et al. 1999), suggesting that *NUP98* may be a very frequent target for therapy-related, but also for de novo forms of chromosomal translocations causing not only AML and MDS, but also acute lymphocytic leukemia (ALL; Hussey et al. 1999) and chronic myelomonocytic leukemia (CML; Wong et al. 1999).

Another vertebrate nucleoporin implicated in chromosomal translocations associated with AML is CAN/Nup214 (von Lindern et al. 1992). CAN/Nup214, a putative oncogene product originally identified in human (Snow et al.1987) and rat (Kraemer et al. 1994). CAN/Nup214 represents a constituent of the cytoplasmic fibrils of the NPC (see Fig. 3; Panté et al. 1994), and it forms a complex with Nup88/Nup84 and probably also with p62 (Bastos et al. 1997; Fornerod et al. 1997). Knockout mice carrying a disrupted allele of CAN/Nup214 revealed a lethal phenotype of the null embryos, with defects in NLS-mediated protein import and poly(A)$^+$ RNA export (van Deursen et al. 1996). More specifically, CAN/Nup214 is the target for two distinct export factors, CRM1 and TAP (Fornerod et al.1997; Grüter et al. 1998; Braun et al. 1999; Katahira et al. 1999). CRM1 is the export receptor for proteins that harbor a nuclear export signal (NES) within their amino acid sequence (see Fornerod and Ohno, this Vol.), whereas TAP plays a role in the export of viral RNAs bearing a constitutive transport element (CTE) and is predicted to be involved in the export of mRNA (see Izaurralde, this Vol.). Most likely, CAN/Nup214 is involved in both, the initial step of nuclear protein import, as well as in the terminal step of distinct protein and RNA export pathways.

Evidently, Nup98 and CAN/Nup214 are not the only nucleoporins that are related to human diseases. Tpr (translocated promoter region) was originally identified in oncogenic fusions in combination with the proto-oncogenes *met*, *trk* and *raf* (Mitchell and Cooper 1992). Interestingly, Nup155, a vertebrate nucleoporin of unknown function, might be implicated in mental and developmental retardation, as suggested from the genomic location of the human *NUP155* gene on chromosome band 5p13 (Zhang et al. 1999). Moreover, p62 and the integral membrane protein gp210 (see Table 1 and Fig. 3) apparently play some role in the autoimmune disease primary biliary cirrhosis (PBC; reviewed in Courvalin and Worman 1997). In fact, 25% of PBC patients produce auto-antibodies directed against gp210, predominantly against a distinct 15-amino-acid segment residing within the cytoplasmic, C-terminal-end domain of gp210 (Nickowitz and Worman 1993; reviewed in Courvalin and Worman 1997). Some of these PBC patients also harbor auto-antibodies directed against p62, one of the first-identified and best-characterized vertebrate nucleoporins (reviewed in Panté and Aebi 1996a; Kinoshita et al. 1999; Stoffler et al. 1999a), as well as against the newly identified potential human nucleoporin PBC68 (Theodoropoulos et al. 1999).

gp210 is one of the two transmembrane nucleoporins that have been identified in vertebrates, the other being POM121. gp210 was first isolated and identified in rat and *Drosophila* (Gerace et al. 1982; Filson et al. 1985). It appears to be part of the NPC's central framework (Fig. 3; Greber et al. 1990), with its large N-terminal domain(95% of its total mass) protruding into the lumen of the double-membraned NE (Wozniak and Blobel 1992). A short, 21-residue-long transmembrane segment is sufficient to target gp210 to the pore membrane (Wozniak and Blobel 1992). It has been suggested that gp210 anchors the NPC in the pore membrane and thereby exerts a topogenic role in membrane folding during NPC formation (Greber et al. 1990; Jarnik and Aebi 1991; Gerace 1992; Hinshaw et al. 1992). POM121, which was originally identified in rat (Hallberg et al. 1993), consists of a short N-terminal and a long FG-repeat containing C-terminal domain, with a short transmembrane segment joining the two distinct end domains (Hallberg et al. 1993). The short N-terminal domain of POM121 resides in the lumen of the NE, whereas the long C-terminal domain forms part of the central framework of the NPC (Söderqvist and Hallberg 1994). Like gp210, POM121 is thought to act as a membrane anchor of the NPC (Hallberg et al. 1993).

4 What Is the Functional Significance of the Multiple Locations of Nucleoporins?

Immunogold-electron microscopy has localized p62, one of the first nucleoporins identified and characterized in vertebrates (Starr et al. 1990; Carmo-Fonseca et al. 1991; Cordes et al. 1991), to three distinct NPC sites: (1) near or at the cytoplasmic entry to the central pore; (2) near or at the nuclear entry to the central pore; and (3) near or at the distal ring of the nuclear basket (see Fig. 3; Panté and Aebi 1993; Guan et al. 1995). At the cytoplasmic and nuclear entry to the central pore, p62 is in complex with three other nucleoporins, p58, p54 and p45 (see Fig. 3; reviewed in Panté and Aebi 1996a). In addition, p62 forms a complex with Nup93 that locates (1) to the nuclear periphery of the central pore, and (2) to the distal ring of the nuclear basket (see Fig. 3; Grandi et al. 1997). The putative yeast homologue of p62, Nsp1p, forms a complex with the three yeast nucleoporins Nup57p, Nup49p and Nic96p (Grandi et al. 1995; see also Schlaich et al.1997). Similar to vertebrate p62, yeast Nsp1p and its interacting nucleoporins exhibit multiple locations at the cytoplasmic and the nuclear periphery of the NPC (see Fahrenkrog et al. 1998, 2000; Rout et al. 2000). The multiple locations of nucleoporins suggest that these may represent mobile - rather than stationary - constituents of the NPC. More specifically, nucleoporins such as p62 or Nsp1p may actually shuttle between distinct NPC sites, most likely in association with the cargo complex via a direct interaction with, for example, importinβ (Stochaj et al. 1998). Moreover, the quasi-symmetrical distribution of many nucleoporins relative to the central plane of the NPC (see Fahrenkrog et al. 1998; Stoffler et al. 1999a; Rout et al. 2000) is to be expected from the high degree of 822 symmetry of the basic framework of the NPC, as well as from the structural similarity of the cytoplasmic and nuclear ring moieties of vertebrate NPCs (see Figs. 1 and 2). Also, it is conceivable that the same nucleoporins might serve as docking or parking sites for both import and export cargo at the cytoplasmic and the nuclear periphery of the central pore, thereby funneling cargo into the central pore from either the cytoplasmic or the nuclear compartment (see Rout et al. 2000).

Shuttling of nucleoporins has in fact recently been demonstrated for Nup98, Nup153 and CAN/Nup214(Boer et al. 1997; Nakielny et al. 1999; Zolotukhin and Felber 1999). For example, upon addition of the transcription inhibitor actinomycin D, Nup98 translocates from the nucleus to the cytoplasm of HeLa cells (Zolotukhin and Felber 1999). Moreover, Nup153, a constituent of the nuclear basket involved in mRNA export (Bastos et al. 1996; Shah et al. 1998), shuttles from the nuclear periphery of the NPC to its cytoplasmic face in association with the export cargo complex (Nakielny et al. 1999). Also, overexpression of CAN/Nup214 in HeLa cells causes its mislocation from the cytoplasmic fibrils to the nuclear basket (Boer et al. 1997), indicating that the location of a nucleoporin might not only change during nucleocytoplasmic transport, but may also depend on the expression level within the cell.

During mitosis the NE and the NPCs disassemble and become dispersed throughout the mitotic cytoplasm (reviewed in Gant and Wilson 1997; Bodoor et al.1999b). Hence this represents another process during which nucleoporins become mobile. In fact, no NPC or NPC-like structures can be detected during mitosis, but distinct nucleoporin subcomplexes, for example, the p62 complex (i.e. p62 in complex with p58, p54 and p45; see above), CAN/Nup214-Nup84, CAN/Nup214-p62, Nup98/Gle2, as well as Nup153 homo-oligomers (Finlay et al. 1991; Bodoor et al. 1999a; Matsuoka et al. 1999) are observed. As might be expected, NPC disassembly is regulated by phosphorylation(Macaulay et al. 1995; Favreau et al. 1996; Collas 1998). Reassembly of NPCs starts at anaphase and proceeds through early G1 (reviewed in Gant and Wilson 1997; Bodoor et al. 1999b). During this process NPCs are inserted into areas of intact nuclear membranes (Macaulay and Forbes 1996; Goldberg et al. 1997), i.e. via formation of small pores in the outer nuclear membrane, termed dimples, as demonstrated by field-emission in-lens scanning electron microscopy (FEISEM; Goldberg et al. 1997; reviewed in Gant et al. 1998). Upon distinct intermediate steps, reassembly is completed by the formation of the peripheral NPC structures, for example, the cytoplasmic fibrils and the nuclear basket (Goldberg et al. 1997; reviewed in Gant et al. 1998). During reformation of NPCs distinct nucleoporins are recruited stepwise (Bodoor et al. 1999b; Haraguchi et al. 2000). For example, Nup153 and POM121 associate with the chromatin during late anaphase (Bodoor et al. 1999a). Whereas the initial Nup153-chromatin association appears to be membrane-independent, POM121 only associates after membrane formation (Bodoor et al. 1999a). During telophase, the p62 complex, CAN/Nup214, Nup84 and RanBP2/Nup358 are recruited (Bodoor et al. 1999a). Finally, at late telophase/early G1, association of gp210 and Tpr with the NPC occurs after the nuclear import activity of the newly assembled nucleus is recovered (Bodoor et al. 1999b; Haraguchi et al. 2000). These recent reassembly studies (see Bodoor et al., 1999a,b; Haraguchi et al. 2000) suggest that it is POM121, rather than gp210, that specifies the initial assembly site for the formation of new NPCs after mitosis.

5 From Nuclear Pore Complex Structure to Nucleocytoplasmic Transport

Signal-dependent/factor-mediated bi-directional nucleocytoplasmic transport of proteins RNAs and RNP particles is thought to occur through the central, ~45-nm diameter pore of the NPC. In contrast, passive diffusion of small molecules and ions might take place via eight distinct peripheral channels exhibiting an average physical diameter of about 10 nm (see Figs. 1 and 2). Whereas it is without any doubt that the central pore represents the sole channel through which active transport processes occur (see Kiseleva et al. 1996; Panté and Aebi 1996b; Panté et al. 1997), the functional significance of the eight peripheral channels has remained controversial.

To this end, microinjection of polyethylene glycol (PEG)-coated colloidal gold particles into either the cytoplasm or the nucleus of *Xenopus* oocytes has delineated a single transport gate midway in the central channel that constricts passive diffusion (Feldherr and Akin 1997), without, however, excluding the possibility that passive diffusion might also occur via the eight peripheral channels. Following the path of fluorescently labeled dextran particles in isolated *Xenopus* nuclei by a recently established fluorescence microscopic method, termed optical single-transporter recording (OSTR), also revealed passive diffusion to take place via a single channel located in the center of the NPC, rather than through multiple channels (Keminer et al. 1999; Keminer and Peters 1999). In contrast, electrical conductance measurements across the nuclear envelope of *Xenopus* oocyte nuclei by a "nuclear hourglass" technique revealed that the peripheral NPC channels exhibit a high ionic permeability independent from the active protein transport mechanism (Danker et al. 1999). Taken together, the role of the eight peripheral NPC channels and the "mechanism" of passive diffusion remain to be established.

A recent microinjection study in HL60 cells and rat liver nuclei has mapped the nuclear export path through NPCs of immunogold-labeled RNAs and RNP particles (Iborra et al. 2000). Surprisingly, no RNA was detected near the NPCs' central axis on its way to the central pore at the nuclear membrane. Instead, the label was distributed around the sides, apparently entering the nuclear basket near or at the inner nuclear membrane. In addition, many NPCs appeared to be associated with particular transport factors and/or cargoes to the exclusion of others. Some of the electron micrographs also suggested that NPCs have different functional zones, for example, where SR proteins are dephosphorylated, and where hnRNP C is removed from mRNA. Unfortunately, the electron microscopy data from which these conclusions were drawn become somewhat ambiguous when viewed kinetically, and they are definitely in contrast to previous electron microscopy data presented by several workers in the field (see Kiseleva et al. 1996; Panté and Aebi 1996b; Panté et al. 1997). Hence, functional interpretation of these recent electron microscopy data should be made with caution.

Aiming at a more structure-based functional understanding of nucleocytoplasmic transport, a number of recent studies have attempted identification and characterization of distinct functional states of the NPC and to correlate these with different morphological appearances of the NPC. To this end, it has been known for some time that structural changes of the NPC can be induced by physiological effectors such as calcium or nucleotides (see Jarnik and Aebi 1991). Moreover, during oogenesis, NTF2/p10, the nuclear import receptor for RanGDP, appears to regulate nucleocytoplasmic transport by modulating the functional diameter of the central channel of the NPC (Feldherr et al. 1998). Toward this goal, atomic force microscopy (AFM) has become the method of choice to monitor structural changes within individual NPCs (Fig. 4; see Rakowska et al. 1998; Stoffler et al. 1999b; Wang and Clapham 1999). Most significant in this endeavor, AFM has now evolved to the point where it has become possible to record the surface topography of single particles, such as, for example, NPCs, in a physiological buffer environment at molecular detail, a prerequisite to enable direct correlation of structural changes with distinct functional states (see Engel et al. 1999; Stoffler et al. 1999b; Wang and Clapham 1999).

As a first example, the effect of ATP on NPC conformation was explored with isolated NEs from *Xenopus* oocytes and with rat cardiomyocytes (Rakowska et al. 1998; Perez-Terzic et al. 1999). Accordingly, addition of ATP to the buffer solution caused dramatic conformational changes of the NPC. More specifically, upon addition of ATP to isolated *Xenopus* oocyte NEs, the height of the NPCs protruding from the cytoplasmic face of the NE increased significantly with a concomitant decrease of the diameter of the NPCs' central pore (Rakowska et al. 1998). In contrast, in cardiomyocytes, depletion of ATP reduced the opening of the central pore with an overall radial expansion of the cardiac NPC, however, without affecting the height of the NPC (Perez-Terzic et al. 1999). Similarly, depletion of Ca^{2+} induced the closure of the cytoplasmic ring of the NPC in cardiomyocytes concurrent with a reduction of the height of the NPC (Perez-Terzic et al. 1999).

Importantly, to improve the adhesion of the sample to the specimen support, the material used in these studies was chemically fixed, detergent-extracted, and/or dehydrated and rehydrated during sample preparation. Hence, it cannot be ruled out that some of the observed structural changes were due to specimen preparation effects rather than reflecting bona fide distinct functional states. To address this question more systematically, time-lapse AFM was carried out under "fully native" conditions, i.e. without employing chemical fixatives, detergent treatment or dehydration (Fig. 4; Stoffler et al.1999b; Wang and Clapham 1999). Under these conditions, the repeated opening and closing

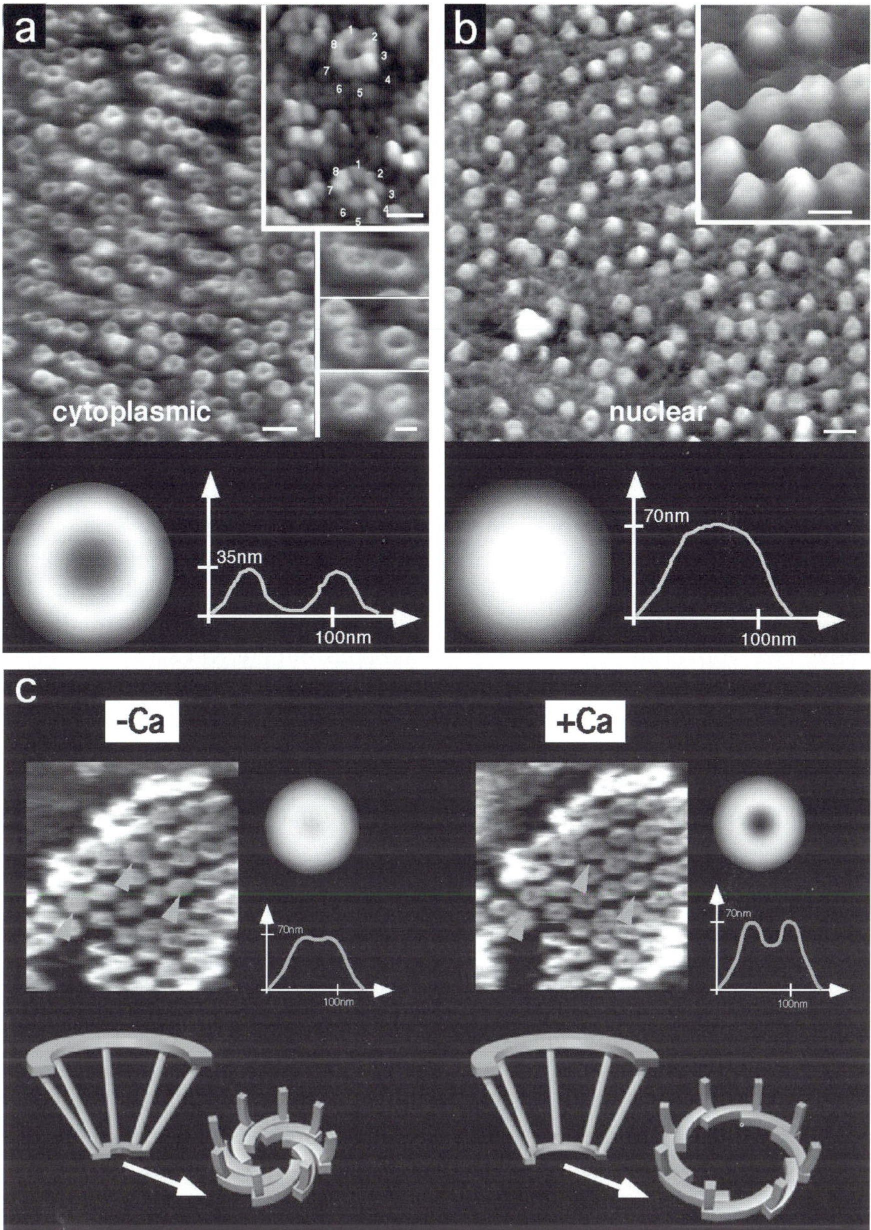

Fig. 4a–c. Atomic force microscopy (AFM) of native *Xenopus* NPCs. **a** The cytoplasmic face of the NPC appears doughnut-like, with an eight-fold rotational symmetry of the central framework. **b** The nuclear face of the NPC appears dome-like; remnants of the nuclear lamina can be seen. **c** Direct visualization by time-lapse AFM of the reversible calcium-mediated opening (+100 μM Ca^{2+}) and closing (+1 mM EGTA) of the nuclear basket, i.e. via its distal ring which may act as an iris-like diaphragm. *Scale bars* 200 nm; *insets* 100 nm

of the nuclear basket of individual NPCs could be observed upon addition and removal of Ca^{2+} (Fig. 4c; Stoffler et al. 1999b). More specifically, the distal ring of the nuclear basket opens 20–30 nm without affecting the overall height and shape of the basket, suggesting that it might act as an "iris-like" diaphragm (Panté and Aebi 1996a), closing in the absence of Ca^{2+} and opening upon addition of Ca^{2+}. Moreover, the observed Ca^{2+} effect appears to be limited to the nuclear face of the NPC, as no evidence of a structural change was depicted by time-lapse AFM at its cytoplasmic face. In contrast, Wang and Clapham (1999) reported that the central plug of the NPC shifted toward the cytoplasmic face of the NPC upon Ca^{2+} depletion concurrent with a decrease in the lumen of the central NPC pore. Similarly, movement of the central plug toward the cytoplasm upon depletion of Ca^{2+} was observed in cardiomyocytes under non-native conditions(Perez-Terzic et al. 1999).

In this context, it should be noted that the nature and/or functional significance of the central plug remains a controversial structural component of the NPC. For example, 3-D reconstructions of membrane-detached (i.e. by detergent treatment) NPCs unveiled a massive barrel-shaped particle residing in the central pore of vertebrate (Akey and Radermacher 1993) and yeast (Yang et al. 1998) NPCs. In contrast, the central plug appeared as a solid, highly variable mass exhibiting little substructure in a tomographic 3-D reconstruction of native (see above) thick ice-embedded *Xenopus* oocyte NPCs (Stoffler et al. 2001). A distinct bona fide NPC substructure seated in the central pore has also been suggested from FEISEM studies of *Chironomus* NPCs (Kiseleva et al. 1998). As an alternative interpretation, it appears conceivable that the central plug represents cargo caught in transit (see above) or, caused by sample preparation, it might represent the basket's distal ring having collapsed – or been "squashed" – into the NPC's central pore. Toward resolving this ambiguity, AFM has been employed to directly visualize the central plug, thereby deciding whether it represents a stationary or a mobile NPC moiety. Whereas these studies have demonstrated that the central plug can indeed move, structural interpretation of these observations has been controversial (Perez-Terzic et al. 1999; Stoffler et al. 1999b; Wang and Clapham 1999), so that the functional nature of the central plug remains elusive.

Collectively, the numerous structural and biochemical data that have been gathered on the NPC and on nucleocytoplasmic transport, while far from being complete, point to a rather simple translocation mechanism, for example, one being driven by confined diffusion along a gradient (i.e. both in terms of number and affinity) of binding sites in the direction of cargo movement (Fig. 5). As a first step, a nuclear import cargo docks to the cytoplasmic fibrils of the NPC by its receptor, i.e. to an excess of relatively low-affinity nucleoporin sites (e.g., to FG repeats residing on CAN/Nup214; see Fig. 3). Next, the cargo complex is "handed over" (e.g. by passive bending of the cytoplasmic fibrils) to higher-affinity sites located near or at the cytoplasmic entry to the central pore (e.g. the p62 complex; see Fig. 3), for example, acting as an "affinity gate" (see Rout et al.

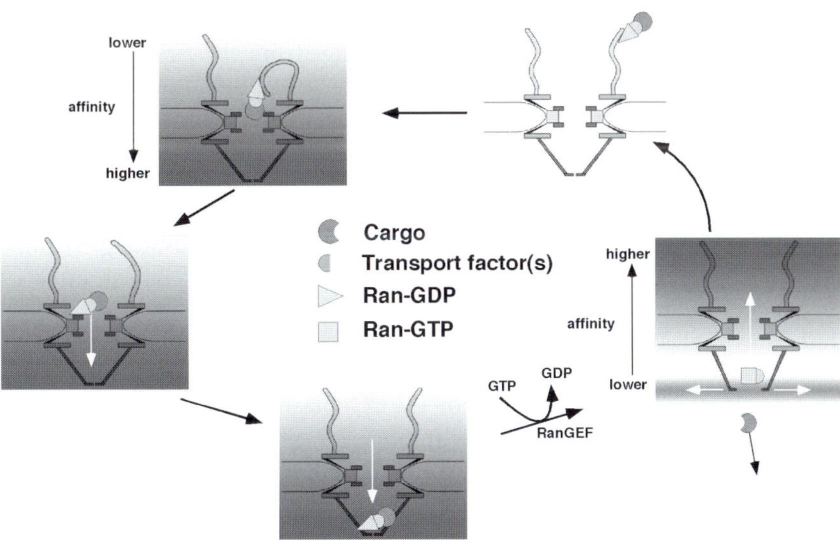

Fig. 5. A possible mechanistic model, involving affinity gating, that describes signal-mediated transport of cargoes through the NPC

2000). From there the cargo complex is dispatched into the central pore and "pulled" into the nuclear basket by an excess of relatively high-affinity binding sites lining the nuclear periphery of the NPC. Once the cargo complex has reached its terminal binding site which, most likely, involves the FG repeats of Nup153 residing at the distal ring of the nuclear basket (see Fig. 3), RanGDP bound to the import receptor is converted to RanGTP by the action of the nucleotide exchange factor RanGEF, thereby acting as a molecular switch causing release of the cargo into the nucleus concurrent with unbinding of the receptor-RanGTP complex for recycling into the cytoplasm (see Fig. 5 and below). To increase diffusion of the cargo into the nucleus, similar to an iris diaphragm (see Fig. 4c), the distal ring may dilate, possibly induced by divalent cations such as Ca^{2+} (see above; Stoffler et al. 1999b). In fact, the maximum opening diameter of the distal ring (~25–30 nm) corresponds well to the exclusion size for particles that can translocate through the NPC (Feldherr et al. 1984).

Nuclear export cargoes, as a first step on route to the cytoplasm, dock to nucleoporins residing at the nuclear basket (e.g., Nup153; see Fig. 3) via their export receptors and in the presence of RanGTP. Similar to import cargoes (see above), driven by a gradient of binding sites, nuclear export cargoes, in complex with an export receptor, eventually reach a terminal binding site most likely residing within the cytoplasmic fibrils and involving the nucleoporin RanBP2/Nup358 (see Fig. 3). Catalyzed by RanBP1 and RanGAP, the export cargo complex is released from the cytoplasmic fibrils by the hydrolysis of RanGTP to RanGDP (i.e. Ran acting again as a molecular switch; see above) and, concurrently, the export cargo dissociates from the export receptor and diffuses into the cytoplasm. Similar to export receptors, import receptors in complex with RanGTP (see above) also have a higher affinity for cytoplasmic nucleoporins, hence they shuttle back to the cytoplasmic side of the NPC (see Fig. 5).

To maintain a large nuclear RanGTP pool, not only is RanGDP imported into the nucleus in complex with import cargo (see above and Fig. 5), but it is also re-imported by its own import receptor, NTF2/p10 (Ribbeck et al. 1998; Smith et al. 1998), and recharged with GTP by the action of RanGEF (Bischoff and Ponstingl 1991a,b). The directionality of signal-mediated transport of cargoes across the nuclear envelope is regulated (1) by the asymmetric distribution of particular nucleoporins on the cytoplasmic or the nuclear face of the NPC with distinct affinity to either import or export receptors, and (2) by the asymmetric distribution of the small GTPase Ran, presumably with RanGDP predominantly residing in the cytoplasm and RanGTP being in excess in the nucleus.

6 Concluding Remarks

Over the past several years, the 3-D structural organization of the *Xenopus* NPC has been extensively investigated by electron microscopy, so that slowly but definitively its substructure and molecular architecture are unfolding. Although identified and/or resolved by 3-D reconstruction, the functional role of some distinct NPC components in nucleocytoplasmic transport, for example, the central plug or the peripheral channels, has remained controversial. Most significantly, time lapse AFM has begun to identify and/or define structurally distinct functional states of the NPC directly correlating with nucleocytoplasmic transport. Whereas evidently all nucleoporins have now been identified and mapped within the yeast NPC, a large fraction of the vertebrate nucleoporins remain to be identified and localized within the 3-D architecture of the NPC. Surprisingly, most of these nucleoporins, i.e. as defined by the location of their end termini or individual epitopes, appear to reside on either the cytoplasmic or the nuclear periphery of the NPC, so that the bulk of the NPC mass has remained a "black hole" in terms of nucleoporins. Atomic structure determination of individual nucleoporins or distinct nucleoporin complexes in combination with more extensive nucleoporin epitope labeling is required to eventually fill this gap. Last but not least, the involvement of distinct nucleoporins in, e.g. leukemias and autoimmune diseases, indicates how important it is to identify and locate the missing nucleoporins. It is conceivable that some of these nucleoporins might in fact be associated with diseases that are caused by disorders in their respective genes. Taken together, although our understanding of NPC structure has grown immensely over the past several years, we still have to go a long way before we can more completely and systematically correlate the 3-D molecular architecture of the NPC with its functional involvement in nucleocytoplasmic transport.

Acknowledgments. We thank Drs. Daniel Stoffler and Bernhard Feja for providing Fig. 2, Daniel Stoffler for providing Fig. 4, and Robert Wyss for his help with Figs. 3 and 5. This work was supported by grants from the European Molecular Biology Organization given to Elisa Izaurralde, the Swiss National Science Foundation, the Human Frontier Science Program (HFSP), and by the Kanton Basel-Stadt and the M.E. Müller Foundation of Switzerland.

References

Ahuja HG, Felix CA, Aplan PD (1999) The t(11;20)(p15;q11) chromosomal translocation associated with therapy-related myelodysplastic syndrome results in an *NUP98-TOP1* fusion. Blood 94:3258–3261

Akey CW, Radermacher M(1993) Architecture of the *Xenopus* nuclear pore complex revealed by 3-dimensional cryo-electron microscopy. J Cell Biol 122:1–19

Allen TD, Cronshaw JM, Bagley S, Kiseleva E, Goldberg MW (2000) The nuclear pore complex: mediator of translocation between nucleus and cytoplasm. J Cell Sci 113:1651–1659

Arai Y, Hosoda F, Kobayashi H, Arai K, Hayashi Y, Nanao K, Kaneko Y, Ohki M(1997) The inv(11)(p15q22) chromosome translocation of de novo and therapy-related myeloid malignancies results in fusion of the nucleoporin gene NUP98, with the putative RNA helicase gene, DDX10. Blood 89:3936–3944

Bangs P, Burke B, Powers C, Craig R, Purohit A, Doxsey S (1998) Functional analysis of Tpr: identification of nuclear pore complex association and nuclear localization domain and a role in mRNA export. J Cell Biol 143:1801–1812

Bastos R, Lin A, Enarson M, Burke B (1996) Targeting and function of nuclear pore complex protein Nup153. J Cell Biol 134:1141–1156

Bastos R, de Pouplana LR, Enarson M, Bodoor K, Burke B (1997) Nup84, a novel nucleoporin that is associated with CAN/Nup214 on the cytoplasmic face of the nuclear pore complex. J Cell Biol 137:989–1000

Bischoff FR, Ponstingl H (1991a) Catalysis of guanine nucleotide exchange on Ran by the mitotic regulator RCC1. Nature 354:80–82

Bischoff FR, Ponstingl H (1991b) Mitotic regulator protein RCC1 is complexed with a nuclear ras-related polypeptide. Proc Natl Acad Sci USA 88:10830–10834

Boer JM, van Deursen JMA, Huib HC, Fransen JAM, Grosveld GC (1997) The nucleoporin CAN/Nup214 binds to both the cytoplasmic and the nucleoplasmic sides of the nuclear pore complex in overexpressing cells. Exp Cell Res 232:182–185

Bodoor K, Shaikh S, Salina D, Raharjo WH, Bastos R, Lohka M, Burke B(1999a) Sequential recruitment of NPC proteins to the nuclear periphery at the end of mitosis. J Cell Sci 112:2253–2264

Bodoor K, Shaikh S, Enarson P, Chowdhury S, Salina D, Raharjo WH, Burke B(1999b) Function and assembly of nuclear pore complex proteins. Biochem Cell Biol 77:321–329

Borrow J, Shearman AM, Stanton VP Jr, Becher R, Collins T, Williams AJ, Dubé I, Katz F, Morris C, Ohyashiki K, Toyama K, Rowley J, Housman DE(1996) The t(7;11)(p15;p15) translocation in acute myeloid leukemia fuses the genes for nucleoporin *NUP98* and class I homeoprotein *HOXA9*. Nat Genet 12:159–167

Braun IC, Rohrbach E, Schmitt C, Izaurralde E (1999) TAP binds to the constitutive transport element (CTE) through a novel RNA-binding motif that is sufficient to promote CTE-dependent RNA export from the nucleus. EMBO J 18:1953–1965

Carmo-Fonseca M, Kern H, Hurt EC (1991) Human nucleoporin p62 and the essential yeast nuclear pore protein NSP1 show sequence similarity and similar domain organization. Eur J Cell Biol 55:17–30

Collas P (1998) Nuclear envelope disassembly in mitotic extracts requires functional nuclear pores and a nuclear lamina. J Cell Sci 111:1293–1303

Cordes VC, Waizenegger I, Krohne G (1991) Nuclear pore complex glycoprotein p62 of Xenopus laevis and mouse: cDNA cloning identification of its glycosylation region. Eur J Cell Biol 55:31–47

Cordes VC, Reidenbach S, Rackwitz HR, Franke WW (1997) Identification of protein p270/Tpr as a constitutive component of the nuclear pore complex-attached intranuclear filaments. J Cell Biol 136:515–529

Cordes VC, Hase ME, Müller L (1998) Molecular segments of protein Tpr that confer nuclear targeting and association with the nuclear pore complex. Exp Cell Res 245:43–56

Courvalin JC, Worman HJ (1997) Nuclear envelope protein autoantibodies in primary biliary cirrhosis. Semin Liver Dis 17:79–90

Danker T, Schillers H, Storck J, Shahin V, Krämer B, Wilhelmi M, Oberleithner H (1999) Nuclear hourglass technique: an approach that detects electrically open nuclear pores in *Xenopus laevis* oocyte. Proc Natl Acad Sci USA 96:13530–13535

Davis LI (1995) The nuclear pore complex. Annu Rev Biochem 64:865–896

Dockendorff TC, Heath CV, Goldstein Al, Snay CA, Cole CN (1997) C-terminal truncations of the yeast nucleoporin Nup145p produce a rapid temperature-conditional mRNA export defect and alterations to nuclear structure. Mol Cell Biol 17:906–920

Emtage JLT, Bucci M, Watkins JL, Wente SR (1997) Defining the essential functional regions of the nucleoporin Nup145p. J Cell Sci 119:911–925

Enarson P, Enarson M, Bastos R, Burke B (1998) Amino-terminal sequences that direct nucleoporin Nup153 to the inner surface of the nuclear envelope. Chromosoma 107:228–236

Engel A, Lyubchenko Y, Müller D (1999) Atomic force microscopy: a powerful tool to observe biomolecules at work. Trends Cell Biol 9:77–80

Fahrenkrog B, Hurt EC, Aebi U, Panté N (1998) Molecular architecture of the yeast nuclear pore complex: localization of Nsp1p subcomplexes. J Cell Biol 143:577–588

Fahrenkrog B, Aris JP, Hurt EC, Panté N, Aebi U (2000) Comparative spatial localization of protein A tagged and endogenous yeast nuclear pore complex proteins by immunoelectron microscopy. J Struct Biol 129:295–305

Fahrenkrog B, Stoffler D, Aebi U (2001) Nuclear pore complex architecture and functional dynamics. Curr Top Microbiol 259:95–117

Favreau C, Worman HJ, Wozniak RW, Frappier T, Courvalin JC (1996) Cell cycle-dependent phosphorylation of nucleoporins and nuclear pore membrane protein gp210. Biochemistry 35:8035–8044

Feldherr C, Akin D, Moore MS (1998) The nuclear import factor p10 regulates the functional size of the nuclear pore complex during oogenesis. J Cell Sci 111:1889–1896

Feldherr CM, Akin D (1997) The location of the transport gate in the nuclear pore complex. J Cell Sci 110:3065–3070

Feldherr CM, Kallenbach E, Schultz N (1984) Movement of a karyophilic protein through the nuclear pore of oocytes. J Cell Biol 107:1289–1297

Filson AJ, Lewis A, Blobel G, Fisher PA (1985) Monoclonal antibodies prepared against the major *Drosophila* nuclear matrix-pore complex-lamina glycoprotein bind specifically to the nuclear envelope in situ. J Biol Chem 260:3164–3172

Finlay DR, Meier E, Bradley P, Horecka J, Forbes DJ (1991) A compex of nuclear pore proteins required for pore function. J Cell Biol 114:169–183

Fischer R, Cordes VC, Franke WW (1997) Sequence analysis of the nuclear pore complex protein in a lower metazoan: nucleoporin p62 of the coelenterate *Hydra vulgaris*. Gene 185:285–293

Fontoura BMA, Blobel G, Matunis MJ (1999) A conserved biogenesis pathway for nucleoporins: proteolytic processing of a 186-kilodalton precursor generates Nup98 and the novel nucleoporin, Nup96. J Cell Biol 144:1097–1112

Fornerod M, van Deursen J, van Baal S, Reynolds A, Davis D, Murti KG, Fransen J, Grosveld G (1997) The human homologue of yeast Crm1 is in a dynamic subcomplex with CAN/Nup214 and a novel nuclear pore component Nup88. EMBO J 16:807–816

Gant TM, Wilson KL (1997) Nuclear assembly. Annu Rev Cell Dev Biol 13:669–695

Gant TM, Goldberg MW, Allen TD (1998) Nuclear envelope and nuclear pore assembly: analysis of assembly intermediates by electron microscopy. Curr Opin Cell Biol 10:409–425

Gerace L, Ottaviano Y, Koch-Kondor C (1982) Identification of a major polypeptide of the nuclear pore complex. J Cell Biol 95:826–837

Gerace L (1992) Molecular trafficking across the nuclear pore. Curr Opin Cell Biol 4:637–645

Gigliotti S, Callaini G, Andone S, Riparbelli MG, Pernas-Alonso R, Hoffmann G, Grazani F, Malva C (1998) Nup154, a new *Drosophila* gene essential for male and female gametogenesis is related to the Nup155 vertebrate nucleoporin gene. J Cell Biol 142:1195–1207

Görlich D, Kutay U (1999) Transport between the cell nucleus and the cytoplasm. Ann Rev Cell Dev Biol 15:607–660

Goldberg MW, Wiese C, Allen TD, Wilson KL (1997) Dimples, pores, star-rings, and thin-rings on growing nuclear envelopes: evidence for structural intermediates in nuclear pore assembly. J Cell Sci 110:409–420

Grandi P, Schlaich N, Takotte H, Hurt EC (1995) Functional integration of Nic96p with a core nucleoporin complex consisting of Nsp1p, Nup49p and a novel protein Nup57p. EMBO J 14:76–87

Grandi P, Dang T, Panté N, Shevchenko A, Mann M, Forbes D, Hurt E (1997) Nup93, a vertebrate homologue of yeast Nic96p, forms a complex with a novel 205-kDa protein and is required for correct nuclear pore assembly. Mol Biol Cell 8:2017–2038

Greber UF, Senior A, Gerace L (1990) A major glycoprotein of the nuclear pore complex is a membrane-spanning polypeptide with a large lumenal domain and a small cytoplasmic tail. EMBO J 9:1495–1502

Grüter P, Taberno C, von Kobbe C, Schmitt C, Saavedra C, Bachi A, Wilm M, Felber BK, Izaurralde E (1998) TAP, the human homolog of Mex67p, mediates CTE-dependent RNA export from the nucleus. Mol Cell 1:649–659

Guan T, Müller S, Kleir G, Panté N, Blevitt JM, Häner M, Paschal B, Aebi U, Gerace L (1995) Structural analysis of the p62 complex, an assembly of O-linked glycoproteins that localizes near the central gated channel of the nuclear pore complex. Mol Cell Biol 6:1591–1603

Hallberg E, Wozniak RW, Blobel G (1993) An integral membrane protein of the pore membrane domain of the nuclear envelope contains a nucleoporin-like region. J Cell Biol 122: 513–521

Haraguchi T, Koujin T, Hayakawa T, Kaneda T, Tsutsumi C, Imamoto N, Akazawa C, Sukegawa J, Yoneda Y, Hiraoka Y (2000) Live fluorescence imaging reveals early recruitment of emerin, LBR, RanBP2, and Nup153 to reforming functional nuclear envelopes. J Cell Sci 113:779–794

Hatano Y, Miura I, Kume M, Miura AB (2000) Translocation (1;11)(q23;p15), a novel simple variant of translocation (7;11)(p15;p15), in a patient with AML (M2) accompanied by non-Hodgkin lymphoma and gastric cancer. Cancer Genet Cytogenet 117:19–23

Hinshaw JE, Carragher BO, Milligan RA (1992) Architecture and design of the nuclear pore complex. Cell 69:1133–1141

Hu T, Gerace L (1998) cDNA cloning and analysis of the expression of nucleoporin p45. Gene 221:245–253

Hussey DJ, Nicola M, Moore S, Peters GB, Dobrovic A (1999) The(4;11)(q21;p15) translocation fuses the *NUP98* and *RAP1GDS1* genes and is recurrent in T-cell acute lymphocytic leukemia. Blood 94:2072–2079

Iborra FJ, Jackson DA, Cook PR (2000) The path of RNA through nuclear pores: apparent entry from the sides into specialized pores. J Cell Sci 113:291–302

Ikeda T, Ikeda K, Sasaki K, Kawakami K, Takahara J (1999) The inv(11)(p15q22) chromosome translocation of therapy-related myelodysplasia with NUP98-DDX10 and DDX10-NUP98 fusion transcripts. Int J Hematol 69:160–164

Izaurralde E, Adam SA (1998) Transport of macromolecules between the nucleus and the cytoplasm. RNA 4:351–364

Jarnik M, Aebi U (1991) Towards a 3-D model of the nuclear pore complex. J Struct Biol 107:291–308

Kasper LH, Brindle PK, Schnabel CA, Pritchard CEJ, Cleary ML, van Deursen JMA (1999) CREB binding protein interacts with nucleoporin-specific FG repeats that activate transcription and mediate NUP98-HOAX9 oncogenicity. Mol Cell Biol 19:764–776

Katahira J, Strässer K, Podtelejnikov A, Mann M, Jung JU, Hurt E (1999) The Mex67p-mediated nuclear mRNA export pathway is conserved from yeast to human. EMBO J 2593–2609

Keminer O, Peters R (1999) Permeability of single nuclear pores. Biophys J 77:217–228

Keminer O, Siebrasse JP, Zerf K, Peters R (1999) Optical recording of signal-mediated protein transport through single nuclear pore complexes. Proc Natl Acad Sci USA 96:11842–11847

Kinoshita H, Omagari K, Whittingham S, Kato Y, Ishibashi H, Sugi K, Yano M, Kohno S, Nakanuma Y, Penner E, Wesierska-Gadek J, Reynoso-Paz S, Gershwin ME, Anderson J, Jois JA, Mackay IR (1999) Autoimmune cholangitis and primary biliary cirrhosis – an autoimmune enigma. Liver 19:122–128

Kiseleva E, Goldberg MW, Daneholt B, Allen TD (1996) RNP export is mediated by structural reorganization of the nuclear pore basket. J Mol Biol 260:304–311

Kiseleva E, Goldberg MW, Allen TD, Akey CW (1998) Active nuclear pore complexes in *Chironomus*: visualization of transporter configurations related to mRNP export. J Cell Sci 111:223–236

Kosova B, Panté N, Rollenhagen C, Podtelejnikov A, Mann M, Aebi U, Hurt E (2000) Mlp2p, a component of nuclear pore attached intranuclear filaments, associates with Nic96p. J Biol Chem 275:343–350

Kraemer D, Wozniak RW, Blobel G, Radu A (1994) The human CAN protein, a putative oncogene product associated with myeloid leukemogenesis, is a nuclear pore complex protein that faces the cytoplasm. Proc Natl Acad Sci USA 91:1519–1523

Kwong YL, Pang A (1999) Low frequency of rearrangements of the homeobox gene HOXA9/t(7;11) in adult acute myeloid leukemia. Genes Chrom Cancer 25:70–74

Macaulay C, Forbes DJ (1996) Assembly of the nuclear pore: biochemically distinct steps revealed with NEM, GTP gamma S, and BAPTA. J Cell Biol 135:5–20

Macaulay C, Meier E, Forbes DJ (1995) Differential mitotic phosphorylation of proteins of the nuclear pore complex. J Biol Chem 270:254–262

Matsuoka Y, Takagi M, Ban T, Miyazaki M, Yamamoto T, Kondo Y, Yoneda Y (1999) Identification and characterization of nuclear pore subcomplexes in mitotic extract of human somatic cells. Biochem Biophys Res Commun 254:417–423

Mattaj IW, Englmeier L (1998) Nucleocytoplasmic transport: the soluble phase. Annu Rev Biochem 67:265–306

McMorrow IM, Bastos R, Horton H, Burke B (1994) Sequence analysis of a cDNA encoding a human nuclear pore complex protein, hNup153. Biochim Biophys Acta 1217: 219–223

Mitchell PJ, Cooper CS (1992) Nucleotide sequence analysis of human tpr cDNA clones. Oncogene 7:383–388

Nakamura T, Largaespada DA, Lee MP, Johnson LA, Ohyashiki K, Toyama K, Chen SJ, Willman CL, Chen IM, Feinberg AP, Jenkins NA, Copeland NG, Shaugnessy JD Jr (1996) Fusion of the nucleoporin gene *NUP98* to *HOXA9* by the chromosome translocation t(7;11)(p15;p15) in human myeloid leukemia. Nat Genet 12:154–158

Nakamura T, Yamazaki Y, Hatano Y, Miura I (1999) *NUP98* is fused to *PMX1* homeobox gene in human acute myelogenous leukemia with chromosome translocation t(1;11)(q23;p15). Blood 94:741–747

Nakielny S, Shaikh S, Burke B, Dreyfuss G (1999) Nup153 is an M9-containing mobile nucleoporin with a novel Ran binding domain. EMBO J 18:1982–1995

Nickowitz RE, Worman HJ (1993) Autoantibodies from patients with primary biliary cirrhosis recognize a restricted region within the cytoplasmic tail of nuclear pore membrane protein gp210. J Exp Med 178:2237–2242

Nishiyama M, Arai Y, Tsunematsu Y, Kobayashi H, Asami K, Yabe M, Kato S, Oda M, Eguchi H, Ohki M, Kaneko Y (1999) 11p15 translocations involving the NUP98 gene in childhood therapy-related acute myeloid leukemia/myelodysplastic syndrome. Genes Chromosomes Cancer 26:215–220

Ohno M, Fornerod M, Mattaj IW (1998) Nucleocytoplasmic transport: the last 200 nanometers. Cell 92:327–336

Panté N, Aebi U (1993) The nuclear pore complex. J Cell Biol 122:977–984

Panté N, Aebi U (1996a) Molecular dissection of the nuclear pore complex. Crit Rev Biochem Mol Biol 31:153–199

Panté N, Aebi U (1996b) Sequential binding of import ligands to distinct nucleopore regions during nuclear import. Science 273:1729–1732

Panté N, Bastos R, McMorrow I, Burke B, Aebi U (1994) Interactions and 3-dimensional localization of a group of nuclear pore complex proteins. J Cell Biol 129:925–937

Panté N, Jarmolowski A, Izaurralde E, Sauder U, Baschong W, Mattaj IW (1997) Visualizing nuclear export of different classes of RNA by electron microscopy. RNA 3:498–513

Panté N, Thomas F, Aebi U, Burke B, Bastos R (2000) Recombinant Nup153 incorporates in vivo into *Xenopus* oocyte nuclear pore complexes. J Struct Biol 129:306–312

Perez-Terzic C, Gacy AM, Bortolon R, Dzeja PP, Puceat M, Jaconi M, Prendergast FG, Terzic A (1999) Structural plasticity of the cardiac nuclear pore complex in response to regulators of nuclear import. Circ Res 84:1292–1301

Powers M, Macaulay C, Masiarz FR, Forbes DJ (1995) Reconstituted nuclei depleted of a vertebrate GLFG nuclear pore protein, p97, import but are defective in nuclear growth and replication. J Cell Biol 128:721–736

Powers M, Forbes DJ, Dahlberg JE, Lund E (1997) The vertebrate GLFG nucleoporin, Nup98, is an essential component of multiple RNA export pathways. J Cell Biol 136:241–250

Radu A, Blobel G, Wozniak RW (1994) Nup107 is a novel nuclear pore complex protein that contains a leucine zipper. J Biol Chem 269:17600–17605

Radu A, Moore MS, Blobel G (1995) The peptide repeat domain of nucleoporin Nup98 functions as docking site in transport across the nuclear pore complex. Cell 81:215–222

Rakowska A, Danker T, Schneider SW, Oberleithner H (1998) ATP-induced shape changes of nuclear pores visualized with the atomic force microscope. J Membrane Biol 163:129–136

Raza-Egilmez SZ, Jani-Sait SN, Grossi M, Higgins MJ, Shows TB, Aplan PD (1998) NUP98-HOXD13 gene fusion in therapy-related acute myelogenous leukemia. Cancer Res 58:4269–4273

Reichelt R, Holzenberg A, Buhle EL, Jarnik M, Engel A, Aebi U (1990) Correlations between structure and mass distribution of the nuclear pore complex and of distinct pore complex components. J Cell Biol 110:883–894

Rexach M, Blobel G (1995) Protein import into nuclei: association and dissociation reactions involving transport substrate, transport factors and nucleoporins. Cell 83:683–692

Ribbeck K, Lipowsky G, Kent HM, Stewart M, Görlich D (1998) NTF2 mediates nuclear import of Ran. EMBO J 17:6587–6598

Rosenblum JS, Blobel G (1999) Autoproteolysis in nucleoporin biogenesis. Proc Natl Acad Sci USA 96:11370–11375

Rout MP, Blobel G (1993) Isolation of the yeast nuclear pore complex. J Cell Biol 109:2641–2652

Rout MP, Aitchinson JD, Suprapto A, Hjertaas K, Zhao Y, Chait BT (2000) The yeast nuclear pore complex: composition, architecture and transport mechanism. J Cell Biol 148:635–651

Schlaich NL, Häner M, Lustig A, Aebi U, Hurt E (1997) In vitro reconstitution of a heterotrimeric nucleoporin complex consisting of recombinant Nsp1p, Nup49p and Nup57p. Mol Biol Cell 8:33–46

Shah S, Forbes DJ (1998) Separate nuclear import pathways converges on the nucleoporin Nup153 and can be dissected with dominant-negative inhibitors. Curr Biol 8:1376–1386

Shah S, Tugendreich S, Forbes D (1998) Major binding sites for the nuclear import receptor are the integral nucleoporin Nup153 and the adjacent nuclear filament protein Tpr. J Cell Biol 141:31–49

Smith A, Brownawell A, Macara IG (1998) Nuclear import of Ran is mediated by the transport factor NTF2. Curr Biol 18:6805–6815

Snow CM, Senior A, Gerace L (1987) Monoclonal antibodies identify a group of nuclear pore complex glycoproteins. J Cell Biol 104:1143–1156

Söderqvist H, Hallberg E (1994) The large C-terminal domain of the integral pore membrane protein, POM121, is facing the nuclear pore complex. Eur J Cell Biol 64:186–191

Söderqvist H, Imreh G, Kihlmark M, Linnmann C, Ringertz N, Hallberg E(1997) Intracellular distribution of an integral nuclear pore membrane protein fused to green fluorescent protein. Eur J Biochem 250:808–813

Starr CM, D'Onofrio M, Park MK, Hanover JA (1990) Primary sequence and heterologous expression of nuclear pore glycoprotein p62. J Cell Biol 110:1861–1871

Stoffler D, Fahrenkrog B, Aebi U (1999a) The nuclear pore complex: from molecular architecture to functional dynamics. Curr Opin Cell Biol 11:391–401

Stoffler D, Goldie KN, Aebi U (1999b) Calcium-mediated structural changes of native nuclear pore complexes monitored by time-lapse atomic force microscopy. J Mol Biol 287:741–752

Stoffler D, Feja B, Walz J, Typke D, Baumeister W, Aebi U (2001) Novel structural features of native nuclear pore complexes revealed by cryo-electron tomography. (in preparation)

Strambio-de-Castillia C, Blobel G, Rout MP (1999) Proteins connecting the nuclear pore complex with the nuclear interior. J Cell Biol 144:839–855

Sukegawa J, Blobel G (1993) A nuclear pore complex protein that contains zinc finger motifs, binds DNA, and faces the nucleoplasm. Cell 72:29–38

Teixeira MT, Siniossoglou S, Podtelejnikov S, Bénichou JC, Mann M, Dujon B, Hurt E, Fabre E (1997) Two functionally distinct domains generated by in vivo cleavage of nucleoporin Nup145p: a novel biogenesis pathway for nucleoporins. EMBO J 16:5086–5097

Theodoropoulos PA, Polioudaki H, Koulentaki M, Kouroumalis E, Georgatos SD (1999) PBC68: a nuclear pore complex protein that associates reversibly with the mitotic spindle. J Cell Sci 112: 3049–3059

Van Deursen J, Boer J, Kasper L, Grosveld G (1996) G2 arrest and impaired nucleocytoplasmic transport in mouse embryos lacking the proto-oncogene CAN/Nup214. EMBO J 15:5574–5583

Von Lindern M, Fornerod M, van Baal S, Jaegle M, de Wit T, Bujis A, Grosveld G (1992) The translocation (6;9) associated with a specific type of acute myeloid leukemia, results in fusion of two genes, dek and can, and the expression of a chimeric, leukemia-specific dek-can mRNA. Mol Cell Biol 12:1687–1697

Wang H, Clapham DE (1999) Conformational changes of the in situ nuclear pore complex. Biophys J 77:241–247

Wong KF, So CC, Kwong YL (1999) Chronic myelomonocytic leukemia with t(7;11)(p15;p15) and NUP98/HOXA9 fusion. Cancer Genet Cytogenet 155:70–72

Wozniak RW, Blobel G (1992) The single transmembrane segment of gp210 is sufficient for sorting to the pore membrane domain of the nuclear envelope. J Cell Biol 119:2083–2092

Yang Q, Rout MP, Akey CW (1998) 3-dimensional architecture of the isolated yeast nuclear pore complex: functional and evolutionary implications. Mol Cell 1:223–234

Zhang X, Huanming Y, Corydon MJ, Zhang X, Pedersen S, Korenberg JR, Chen XN, Laporte J, Gregersen N, Niebuhr E, Liu G, Bolund L (1999) Localization of a human nucleoporin 155 gene (NUP155) to the 5p13 region and cloning of its cDNA. Genomics 57:144–151

Zimowska G, Aris JP, Paddy MR (1997) A *Drosophila* Tpr protein homologue is localized both in the extrachromosomal channel network and to nuclear pore complexes. J Cell Sci 110:927–944

Zolotukhin A, Felber BK (1999) Nucleoporins Nup98 and Nup214 participate in nuclear export of human immunodeficiency virus type 1 rev. J Virol 73:120–127

How Ran Is Regulated

F. Ralf Bischoff[1], Klaus Scheffzek[2], and Herwig Ponstingl[1]

1 Ran Belongs to the Superfamily of Ras-Related Proteins

Ran, the *Ras*-related nuclear protein, is a regulatory GTP-binding protein belonging to a large family of small GTPases. Many aspects of cell behavior are controlled by these proteins, including cell shape and cell movement, cell polarity, intracellular transport, and the decision for proliferation or differentiation. Ran is by far the most abundant of these GTPases, and in human HeLa cells it comprises some 0.4% of cellular protein. It regulates import and export of proteins and RNA through the nuclear pores and is involved in the assembly of the mitotic spindle and of the nuclear membrane.

Like molecular switches, these GTPases alternate between a GDP-bound inactive and a GTP-bound active form (Fig. 1). Binding of GTP induces a conformational change in the protein which then can interact with effectors, the direct targets of these GTPases in signal transduction. GTP hydrolysis returns the protein to the inactive form, thereby terminating the transmitted signal. Upon interaction, the functional state of the effector protein changes to trigger a variety of cellular responses. Regulatory proteins modulate the biological activity of the GTP-binding proteins. Guanine-nucleotide-exchange factors promote release of bound GDP, which is then replaced by the more abundant GTP. GTPase-activating proteins accelerate the usually slow rate of GTP hydrolysis by orders of magnitude to return the protein to the resting, GDP-bound state (Bourne et al. 1990, 1991; Boguski and McCormick 1993).

Sequence elements involved in guanine nucleotide binding by the structural module, termed G-domain, are conserved within the superfamily (Fig. 2), whereas regions involved in interactions with effectors remarkably diverge in their amino acid sequences. The residues interacting with the nucleotide are found in different regions of the protein, that line the nucleotide binding pocket. PM-elements 1–3 participate in the binding of the phosphate moieties and in the coordination of the magnesium ion, which is essential for GTP

[1] Division for Molecular Biology of Mitosis, German Cancer Research Center, Im Neuenheimer Feld 280, 69120 Heidelberg, Germany

[2] Structural and Computational Biology Programme, European Molecular Biology Laboratory, Meyerhofstrasse 1, 69117 Heidelberg, Germany

Results and Problems in Cell Differentiation, Vol. 35
K. Weis (Ed.): Nuclear Transport

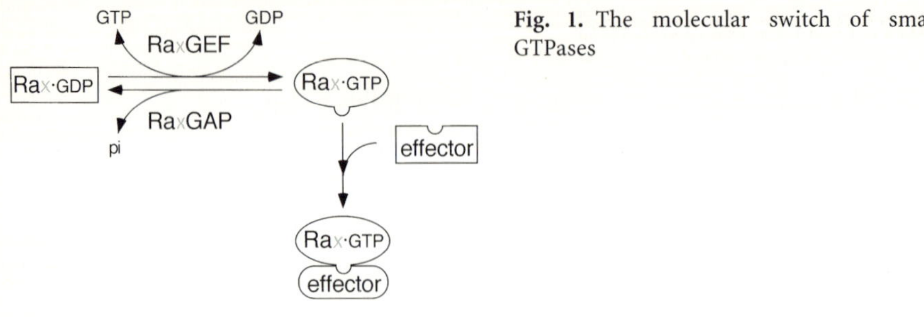

Fig. 1. The molecular switch of small GTPases

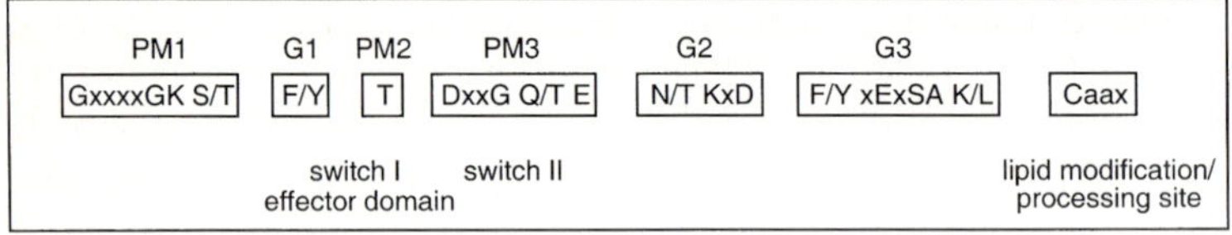

Fig. 2. Sequence elements conserved in small GTPases (according to Valencia et al. 1991). *PM1–3* Phosphate-binding motifs, *G1* guanine-binding motifs, *a* aliphatic amino acid residue, *X* variable residue

hydrolysis. They are located in the P-loop (phosphate-binding loop) and in the switch regions I and II. The switch regions undergo large conformational changes upon GTP binding and hydrolysis. Regions G1–3 stabilize the guanine base by hydrophobic and polar interactions (Fig. 2).

Most of the small GTPases (Valencia et al. 1991) are membrane-associated proteins, anchored by an isoprenoid modification of a C-terminal cysteine. The C-terminal CAAX sequences (cysteine, aliphatic amino acid and X, serine or methionine) are recognized by enzymes attaching farnesyl or geranylgeranyl moieties to the GTPase. To a certain extent this "CAAX-box" also determines the type of the modification. In most cases, the residues C-terminal of the modified cysteine are then removed proteolytically, and the terminal carboxyl group of the cysteine is methylated.

2 Ran

The cDNA sequence of Ran was originally identified as TC4 by screening a human teratocarcinoma cDNA library with degenerate oligonucleotides based on the conserved DTAGQE sequence (part of switch II) of Ras family members (Drivas et al. 1990). The encoded protein was mainly found in the nucleus and was therefore designated Ran (Ras-related nuclear protein; Bischoff and Ponstingl 1991b). As might be expected from its solubility, Ran does not have a signal sequence for conveying membrane anchors; in contrast, an acidic region is found at the C-terminus, which is essential for bind-

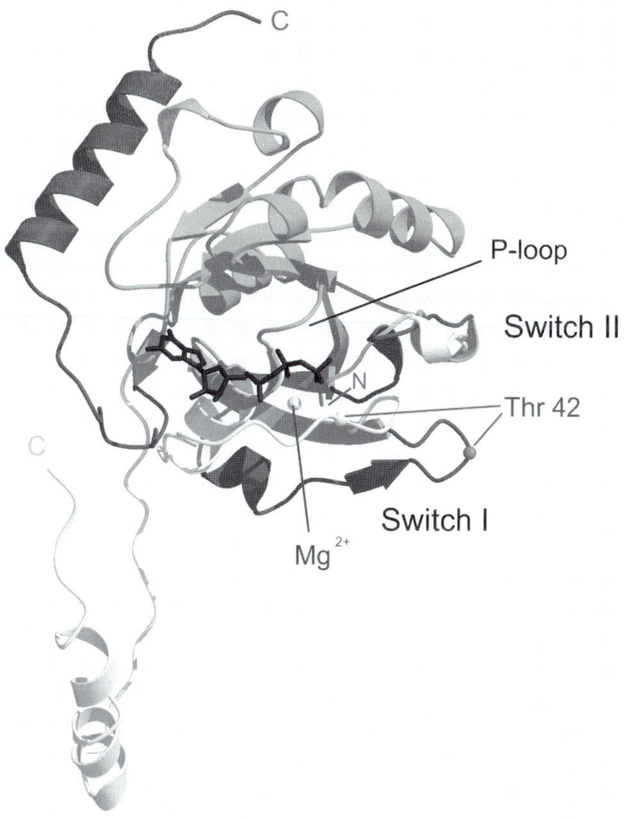

Fig. 3. Structure of Ran in ribbon representation. The G-domain core is shown in *medium gray*. Switch I/II and the C-terminal extension, the segments that change their conformations upon transition between the GDP- and GTP-bound forms, are shown in *dark* (GDP-bound form) and *light gray* (GTP-bound form), respectively. The bound nucleotide is in bond representation. The figure was generated by superimposing the structure of RanGDP on that of RanGppNHp complexed with the Ran binding domain 1 from RanBP2. GppNHp is a nonhydrolyzable GTP analogue

ing a specific class of effectors (see below). Its three-dimensional structure (Scheffzek et al. 1995) is very similar to that of the small GTPase Ras. Five parallel β-strands and one antiparallel strand contribute to a central β-sheet that is surrounded by five α-helices forming the G-domain (Fig. 3), which is conserved in all known GTP-binding proteins. These structural elements are connected by loops, some of which play an essential role in binding and hydrolysis of GTP and in the interactions with effectors. Mutation of Gly-19 to Val (P-loop) or Gln-69 to Leu (switch II) results in proteins unable to hydrolyze bound GTP (Bischoff et al. 1994; Klebe et al. 1995a). Switch I, containing PM2 (Fig. 2), is important for coordination of Mg^{2+}-GTP and for binding of effector proteins. A C-terminal segment emerging from the G-domain forms an extended chain with a terminal α-helix, which in the GDP-bound form is close to the G-domain core.

The switch regions undergo distinct conformational changes on transition from the GTP- to the GDP-bound state (Fig. 3). Remarkably, in the GDP-bound form switch I adopts a conformation that leads to the formation of an additional small β-strand. In the GTP-bound form, as derived from the complex with an effector domain of RanBP2 (Ran binding protein; see below) or with importin-β, a large conformational change occurs bringing the invariant Thr-42 (PM2) into a position for coordination of the bound magnesium ion (Vetter et al. 1999a,b). In addition, the C-terminal segment is extruded and changes its conformation in a "molecular embrace" with the effector, thereby positioning its helical part in a surface groove of RanBP2 (Vetter et al. 1999b).

3 Stimulation of Guanine Nucleotide Exchange by RanGEF

Ran binds GTP specifically and with high affinity ($>10^9\,M^{-1}$; Klebe et al. 1995b). In addition, a low dissociation rate leads to an almost irreversible binding of the guanine nucleotide, the half life of RanGDP and RanGTP complexes being in the range of several hours (Bischoff and Ponstingl 1991b; Klebe et al. 1995a,b). Interestingly, the higher dissociation rate of GTP ($1.1 \times 10^{-4}\,s^{-1}$) than that of GDP ($1.5 \times 10^{-5}\,s^{-1}$ at 25°C) indicates a seven-fold higher affinity of Ran for GDP (Klebe et al. 1995a).

As in other processes regulated by small GTPases, a guanine-nucleotide-exchange factor, RanGEF (previously designated RCC1, "regulator of chromosome condensation"; Nishimoto et al. 1978), specifically binds Ran and stimulates dissociation of the bound nucleotide. The exchange factor in turn is replaced from the intermediary nucleotide-free GTPase-exchange-factor complex by the guanine nucleotide, which is present in the cell in high concentrations. RanGEF was first isolated from HeLa cells as a Ran-RanGEF complex (Bischoff et al. 1990; Bischoff and Ponstingl 1991b).

RanGEF reduces the affinity of Ran for GDP or GTP by five orders of magnitude. The nucleotide exchange rate is increased in the same range (Klebe et al. 1995a). RanGEF does not discriminate between RanGDP and RanGTP, thus the equilibrium of nucleotide binding to Ran is dependent on the cellular concentrations of GDP and GTP and the relative affinities of Ran for each of the nucleotides. The seven-fold lower affinity of Ran for GTP counteracts the conversion to the active GTP-bound state; probably this is compensated for by the withdrawal of RanGTP from the reaction by Ran-specific effectors.

RanGEF has been crystallized and its three-dimensional structure determined by X-ray crystallography (Renault et al. 1998). β-Strands form a "propeller" of seven blades, similar to that observed in the β-subunit of heterotrimeric G-proteins. Mutational analysis of conserved residues identified regions in the molecule that are responsible for the interaction with Ran.

Asp-129, Asp-182 and His-304, all located in close proximity on one side of the structure, were found to be important for the exchange activity (Azuma et al. 1996, 1999). The opposite face comprises the N- and C-terminus and is assumed to bind to chromatin in vivo. RanGEF remains attached to chromatin throughout the cell cycle, and even in mitosis, when in most cells the nuclear envelope breaks down, it appears to signal the position of chromatin to the cell by activating Ran (Carazo-Salas et al. 1999; Ohba et al. 1999; Wilde and Zheng, 1999; Zhang et al. 1999).

Co-crystals of RanGEF and Ran are required to clarify the molecular mechanism of catalyzed nucleotide exchange, but to date no such structures have been published. However, the exchange mechanism has been investigated for the small GTPases Ras and Arf1. Here, a "glutamic acid finger" of the exchange factors Sos and Arno, respectively, interferes with Mg^{2+}- and GDP-β-phosphate binding, thus destabilizing the bound nucleotide (Béraud-Dufour et al. 1998; Boriack-Sjodin et al. 1998; Goldberg, 1998; Mossessova et al. 1998). As the nucleotide binding pocket is not occluded in the complex with the cognate GEFs, rebinding of abundant free GTP is possible.

4 Induction of the Ran GTPase by RanGAP

Intrinsic hydrolysis of Ran-bound GTP is very slow. The half-life of the RanGTP complex is in the range of several hours (Bischoff and Ponstingl 1991a; Bischoff et al. 1994; Klebe et al. 1995b). Regulators have been identified for many Ras-related proteins, which increase their GTPase activity. The purified Ran-specific GTPase activating protein (RanGAP) accelerates hydrolysis of Ran-bound GTP by five orders of magnitude from $1.8 \times 10^{-5}\,s^{-1}$ to $2.1\,s^{-1}$ at 25°C (Bischoff et al. 1994; Klebe et al. 1995a).

RanGAP consists of three domains (Bischoff et al. 1995a). The N-terminal domain, which comprises 385 residues and contains eleven leucine-rich repeats, is responsible for the GAP activity. The three-dimensional structure of this domain has been analyzed for the yeast orthologue Rna1p (Hillig et al. 1999). The repeats form a crescent, the individual repeats each consisting of an α-helix and a β-strand.

There are no sequence homologies to RasGAP, nor are there extensive similarities in the three-dimensional structure. In the interaction between Ras and its GAP, the functionally most important residue, Q61 (Q69 in Ran), is stabilized to position a water molecule that represents the attacking nucleophile (Scheffzek et al. 1998). Most importantly, RasGAP contributes an arginine residue essential for catalysis at the tip of a "finger," and only RasGAP together with Ras forms an efficient GTPase (Scheffzek et al. 1996, 1997; Ahmadian et al. 1997). Arginine introduces a positive charge into the γ-phosphate-binding site of the GTPase. This stabilizes a negative charge that develops in the transition state of the phosphotransfer reaction (GTPase reaction) and stimulates GTP hydrolysis (Scheffzek et al. 1998).

RanGAP activity has been demonstrated for a number of proteins from a wide variety of species. In all these proteins three arginines (R91, 189, and 191, human sequence) are conserved. However, two of these conserved residues (R189 and R191, human sequence) do not appear to affect the activity of *S. cerevisiae* Rna1p (Haberland and Gerke 1999). Mutation of the third (R91 in the human sequence, R74 in *S. pombe*) severely interferes not only with the RanGAP activity, but likewise with Ran binding (Haberland and Gerke 1999; Hillig et al. 1999). Clarification of the catalytic mechanism, however, will have to await determination of the structure of RanGAP in complex with Ran.

The C-terminal domain of RanGAP is separated from the catalytic domain by an acidic region of approximately 40 residues, which is essential for induction of GTP hydrolysis (Haberland et al. 1997). In the homologous yeast proteins (Hopper et al. 1990; Melchior et al. 1993) which also have GAP activity, this C-terminal domain is missing, excluding it as the catalytic domain (Becker et al. 1995; Bischoff et al. 1995a; Corbett et al. 1995). This domain appears to be responsible for covalent modification by the ubiquitin-like protein SUMO-1 (small ubiquitin-like modifier; Matunis et al. 1996; Mahajan et al. 1997), which is found in higher eukaryotes only. This modification of RanGAP is required for its association with RanBP2, a component of the cytoplasmic filaments of the nuclear pore complex (Matunis et al. 1996, 1998; Mahajan et al. 1997). The three-dimensional structure of SUMO-1 is very similar to that of ubiquitin (Bayer et al. 1998). Both proteins display the ββαββαβ scaffold typical for the family of ubiquitin-related proteins. Neither unmodified RanGAP nor SUMO-1 alone bind to RanBP2. Therefore, it is assumed that SUMO-1 induces a conformational change in RanGAP, exposing a binding site for RanBP2 (Bayer et al. 1998).

5 Role of Ran in Protein Import

Interestingly, the antagonistic regulators of Ran, RanGEF and RanGAP, are found on opposite sides of the nuclear envelope. The nucleotide exchange factor is bound to chromatin in the nucleus (Ohtsubo et al. 1989; Bischoff and Ponstingl 1991b). The GTPase activator RanGAP is associated with the cytoplasmic side of the nuclear pore complexes or distributed diffusely in the cytoplasm (Hopper et al. 1990; Melchior et al. 1993; Matunis et al. 1996; Mahajan et al. 1997). Two relevant conclusions regarding Ran function result from these facts: (1) to interact with both factors, Ran has to shuttle as a mobile protein between the two compartments; (2) in the cytoplasm it will be predominantly in the GDP-bound form, in the nucleus RanGTP will prevail.

As detailed in other chapters of this volume, Ran has an essential function in the import of macromolecules into the nucleus and in their export into the cytoplasm. The import and export factors identified so far are Ran-binding proteins. Cytoplasmic cargo proteins destined for the nucleus bind via a nuclear localization signal to one out of a group of importin-β-related import factors of 90–130 kDa which display a faintly similar sequence within an N-terminal Ran-binding domain (Görlich et al. 1995; Moroianu et al. 1995; Chook and Blobel 1999; Cingolani et al. 1999; Vetter et al. 1999a,b).

Upon arrival of a transport complex in the nucleus, RanGTP binding induces a conformational change in the transport factor and thus causes dissociation of the cargo (Rexach and Blobel 1995; Görlich et al. 1996). The empty import factor is then exported with RanGTP attached to it.

This model was corroborated by experiments using permeabilized cells. It was shown that a high concentration of RanGTP in the cytoplasm inhibits import of proteins into the nucleus (Görlich et al. 1996). Only after addition of RanGAP to the assay, transport was continued. This dependency of protein import on functional RanGAP was also demonstrated in vivo in baker's yeast (Corbett et al. 1995). On the other hand, a mutated form of the exchange factor RanGEF can be inactivated in tsBN2 hamster cells by increasing the temperature. At the restrictive temperature the import of microinjected fluorescence-labeled SV40-T-antigen, a protein displaying a classical nuclear localization signal, is blocked (Kadowaki et al. 1993; Tachibana et al. 1994; Dickmanns et al. 1996). The same effect is also observed upon microinjection of RanGAP in high concentrations into the nuclei of *Xenopus* oocytes (Izaurralde et al. 1997). This unequivocally demonstrates that for the continuous import of proteins a high concentration of RanGTP in the nucleus is necessary.

6 Effects of Importin-β-Related Proteins on the Activities of Ran

In keeping with their characteristic Ran-binding, several importins were also identified by an overlay technique, in which proteins are separated by denaturing gel electrophoresis, transferred to nitrocellulose and incubated with Ran[^{32}P]GTP (Lounsbury et al. 1994; Deane et al. 1997; Görlich et al. 1997; Schlenstedt et al. 1997).

Binding of RanGTP to importin-β-related factors results in inhibition of all known enzymatic activities of the GTPase. Intrinsic and RanGAP-induced hydrolysis as well as intrinsic and RanGEF-stimulated exchange of Ran-bound GTP in the complex are inhibited (Floer and Blobel 1996; Görlich et al. 1996; Bischoff and Görlich 1997). This inhibition is thought to result from overlap of the binding sites of importin-β-related factors with the binding sites of the Ran regulators. The RanGTP-importin complexes are very stable. For example, in the presence of RanGAP the RanGTP-importin-β complex dissociates with a half-life of several hours (Bischoff and Görlich 1997; Villa Braslavsky et al. 2000). This resistance to activation of the GTPase was used to determine and compare the affinity of the importin-β-related proteins for RanGTP. For many RanGTP-importin complexes dissociation constants in the order of 1 nM were measured (Floer and Blobel 1996; Görlich et al. 1996, 1997; Deane et al. 1997; Schlenstedt et al. 1997). Interestingly, some importin-β-related transport factors have a very low affinity for RanGTP. These factors appear to be involved in the export of macromolecules from the nucleus (see below).

7 Recycling of Exported Import Factors by RanBP1/RanBP2

The high stability of RanGTP-importin complexes per se would have the disadvantage that neither Ran nor the import factor would be available for further transport events. In search of effectors of the Ran system that specifically bind RanGTP, two unrelated protein families were identified. In addition to importin-β-related factors, two members of the family of RanBP1-related proteins were detected using overlay assays with RanGTP. RanBP1 (Bressan et al. 1991; Coutavas et al. 1993; Bischoff et al. 1995a) is 23kDa and is so far the smallest member of the family. Four Ran-binding domains homologous to it are found in the nuclear pore protein RanBP2, which has a molecular mass of 356kDa (Wu et al. 1995; Yokoyama et al. 1995). In addition, RanBP2 features zinc finger domains, an extended region rich in leucines, numerous FxFG motifs characteristic of nuclear pore proteins, and a C-terminal domain related to cyclophilin with peptidyl-prolyl-isomerase activity. RanBP2 is a filamentous protein of 36nm, an essential constituent of the cytoplasmic fibers of the nuclear pore complex (Delphin et al. 1997).

Isolated RanBP1-homologous domains of RanBP2 behave biochemically similar to RanBP1 (Beddow et al. 1995; Villa Braslavsky et al. 2000). Upon binding of RanGTP they block exchange of Ran-bound nucleotide (Beddow et al. 1995; Bischoff et al. 1995b). However, hydrolysis is not blocked as it is upon binding of importin-β-related proteins, but enhanced by an order of magnitude. RanBP1 thus acts as a GTPase coactivator (Bischoff et al. 1995b; Richards et al. 1995; Schlenstedt et al. 1995). Accessibility of RanBP1-bound RanGTP for RanGAP and the capability of forming heterotrimeric complexes with RanGTP and importin-β-related proteins made RanBP1 a first-choice candidate for a recycling factor that removes RanGTP from the importins. Indeed it was shown for several importin-β-related import factors that their blockage of RanGAP-induced GTP hydrolysis is abolished by RanBP1 (Bischoff and Görlich 1997; Deane et al. 1997; Floer et al. 1997; Görlich et al. 1997; Lounsbury and Macara 1997; Schlenstedt et al. 1997).

The exact mechanism of transport factor release from complexes with RanGTP by binding of RanBP1 or RanBP2 is not yet understood. The acidic C-terminal DEDDDL sequence of Ran appears to be of particular importance in this context. It is unique among all Ras-related proteins. Deletion analyses have shown that it is essential for tight binding of RanBP1 or RanBP2, whereas it impedes the binding of importin-β-related proteins and of RanGAP (Lounsbury et al. 1994). Upon binding to Ran, the complete acidic sequence is wrapped around the RanBP1-homologous domain (Vetter et al. 1999b). Therefore, association of importin with RanGTP is accelerated (Villa Braslavsky et al. 2000). This neutralization of the acidic C-terminus of Ran also explains the long-standing observation that the Ran GTPase is co-activated by RanBP1 (Bischoff et al. 1995b; Richards et al. 1995). On the other hand, binding of importin-β induces a conformational change in Ran which exposes the acidic C-terminal sequence. This has been shown using a monoclonal antibody directed to this region and its recognition only when bound to importin-β (Hieda et al. 1999). Presumably RanBP1 initially binds to the acidic C-terminus of Ran. In this intermediate complex, RanBP1 and importin may compete for Ran binding (Villa Braslavsky et al. 2000). The transport factor may be released from this complex when, in addition, the nucleotide binding domain of Ran is occupied by RanBP1 (Bischoff and Görlich 1997; Villa Braslavsky et al. 2000). To shift the equilibrium of this process towards dissociation, Ran-bound GTP in the RanBP1-RanGTP complex is hydrolyzed upon induction by RanGAP. Alternatively, accessibility of RanGTP for RanGAP in the heterotrimeric RanBP1-RanGTP-importin complex may be enhanced and dissociation would take place after GTP has been hydrolyzed.

The lower affinity of Ran-binding proteins 1 and 2 for GDP-bound Ran largely results from the 1,000-fold higher dissociation rate compared to that of RanGTP, whereas the association rate in both situations is approximately the same (Kuhlmann et al. 1997). Rapid binding of GDP-bound Ran may explain the existence of RanBP1-RanGDP-importin-β complexes, although RanGDP has only a low affinity for the individual factors (Chi et al. 1996; Deane et al. 1997). Importin-β may stabilize RanBP1-bound RanGDP, involving an acidic domain in importin-β, ^{335}DENDDDW^{342}, very similar to the C-terminus of Ran. It is speculated that the RanBP1-RanGDP-importin-β complex plays a role in nuclear import (Richards et al. 1995; Chi et al. 1996).

In higher eukaryotes RanBP2 fulfills the requirements for efficient recycling of RanGTP-transport receptor complexes:

- It is a component of the nuclear pore complex that has to be passed by all proteins during export (Wilken et al. 1995; Wu et al. 1995; Yokoyama et al. 1995; Matunis et al. 1996; Mahajan et al. 1997).
- It displays numerous FxFG motifs and a zinc finger domain, thought to be binding sites for the export complexes, where disassembly could take place (Wu et al. 1995; Yokoyama et al. 1995; Singh et al. 1999).
- It is tightly associated with RanGAP that has been modified by SUMO (Matunis et al. 1996; Mahajan et al. 1997). This is thought to increase the efficiency of the recycling reaction. The importance of RanBP2 and RanGAP for this reaction is illustrated by a *Drosophila* mutant devoid of the enzyme for attaching the SUMO modification to RanGAP and therefore lacking the capacity to form a RanBP2-RanGAP complex. In embryos of that mutant the bicoid-transcription factor required for segmentation can no longer be effectively imported into the nucleus and bicoid-regulated developmental genes are deregulated (Epps and Tanda 1998).

- It provides binding sites for RanGDP-importin which may suffice to keep importins at the nuclear pore complex for the next import event (Chi et al. 1996; Deane et al. 1997).

In higher eukaryotes RanBP1 cannot fully substitute for the function of RanBP2. It rather may be considered a second line of defense against exported RanGTP-importin complexes which have escaped from dissociation at the nuclear pore complex. In baker's yeast, there is no homologue of RanBP2. The recycling of the RanGTP-importin complexes very likely is achieved only by the RanBP1 homologue Yrb1p (Butler and Wolfe 1994; Schlenstedt et al. 1995). Temperature sensitive mutants having a defective *YRB1* gene display impaired transport activity, and deletion of the gene is lethal (Schlenstedt et al. 1995).

For its role in recycling of exported transport complexes, RanBP1 should be confined to the cytoplasm. Surprisingly, however, it also appears to be actively imported into the nucleus. It rapidly accumulates in the nucleus when import factors or the export machinery are defective (Schlenstedt et al. 1997; Hellmuth et al. 1998; Plafker and Macara 2000). Small amounts of nuclear RanBP1 (Yrb1p in yeast) have no effect on transport complexes, presumably because it is rapidly bound to RanGTP and thereby has lost the capacity to dissociate transport complexes (Maurer et al. 2001). However, RanBP1 microinjected in high concentrations inhibits RNA export (Izaurralde et al. 1997), possibly by sequestering all the RanGTP that is required to form the respective export complexes.

8 Export of Macromolecules from the Nucleus

Import and export are coordinated by RanGTP, which in the nucleus is present in high concentrations and in the cytoplasm in low concentrations. For import, macromolecules that have a nuclear localization signal are bound in the cytoplasm by importins and transported to the nucleus through the pores. There, binding of RanGTP to importin results in release of the cargo. Importin is re-exported as a complex with RanGTP to the cytoplasm, where RanBP1 or RanBP2 and RanGAP remove the bound RanGTP. Importin is now ready for an additional round of substrate binding and import.

When proteins are exported from the nucleus, RanGTP has the opposite task. Proteins destined for export have a nuclear export signal (NES) that is recognized by exportins. Like importins, exportins belong to the family of importin-β-related transport factors; yet unlike importins, they generally have a low affinity for RanGTP as well as for their respective export substrates. However, if one of the partners is already bound, it induces a conformational change in the exportin that greatly improves binding of the other partner (Mattaj and Englmeier 1998; Görlich and Kutay 1999). The high RanGTP concentration of some 10 μM in the nucleus greatly favors this co-operative reaction. The resulting RanGTP-exportin-substrate complex is translocated through the nuclear pore and is dissociated in the cytoplasm by RanBP1/2 and

RanGAP. Free exportin can return to the nucleus as monomer and begin an additional export cycle.

As discussed above, RanBP1 appears to be actively imported into the nucleus. It is returned to the cytoplasm by exportin-1, which mainly exports proteins with a nuclear export signal. However, RanBP1 is a very special substrate. It binds to the exportin via RanGTP (Künzler et al. 2000; Maurer et al. 2001), whereas normal export substrates bind to a region of the exportin different from the Ran-binding domain. Correspondingly, the acidic C-terminus of Ran is required for RanBP1 binding (as discussed above for recycling of the export complexes), whereas it is dispensable for binding of normal export substrates. RanBP1 probably is re-exported as a component of this stable ternary recycling complex of RanBP1, RanGTP, and exportin-1, which upon export dissociates in the cytoplasm in the presence of RanGAP (Maurer et al. 2001).

9 Role of RanBP3 in Exportin-1-Mediated Export

The RanBP3 family consists of nuclear proteins with a region homologous to RanBP1 and RanBP2. They are derived from one precursor transcript by differential splicing and have very low affinities ($K_d \approx 100\,\mu M$) for RanGTP (Mueller et al. 1998). The homologue in baker's yeast is the nuclear Yrb2p. Like the human RanBP3 proteins, it has a C-terminal Ran-binding domain and several FxFG motifs that are characteristic for nuclear pore proteins (Noguchi et al. 1997; Taura et al. 1997). The homologue in *S. pombe* is hba1 (Turi et al. 1996). An additional representative of this family of proteins is the nuclear pore protein Nup2p (Loeb et al. 1993). Deletion of *YRB2* is not lethal for yeast, yet cells stop growing at reduced temperature and accumulate export substrates in the nucleus (Taura et al. 1997). This defect can be compensated for by overexpression of the genes encoding the exportin Xpo1p or the yeast homologue of Ran (Noguchi et al. 1999). This and the physical association of the *XPO1* and *YRB2* gene products indicate a joint involvement in the export of proteins from the nucleus. Human RanBP3a and RanBP3b proteins of 60 and 53 kDa bind to the export factor exportin-1 (F.R. Bischoff, U. Kutay, L. Englmeier, unpublished observations). While exportin-1 and RanBP3 each bind RanGTP with very low affinity, their complex has an affinity in the subnanomolar range. The resulting RanBP3-RanGTP-exportin1 complex has a low sensitivity for dissociation induced by RanBP1/RanGAP in comparison to complexes of RanGTP, exportin-1, and export substrates. This indicates that, similar to the Yrb1p-Xpo1p and Yrb2p-Xpo1p complexes in yeast (Maurer et al. 2001), binding of RanBP3 to exportin-1 does not comply with the classical substrate-binding mode (F.R. Bischoff, U. Kutay, L. Englmeier, unpubl. observ.). RanBP3/Yrb2p may favor the binding of export substrate to the RanGTP-exportin-1 complex in the nucleoplasm. The resulting export complex is bound to proteins of the nuclear pore complex, and RanBP3 is released (Taura et al. 1998). Upon translocation into the cytoplasm, the export complex is dissociated by RanBP1 and RanGAP.

10 Import of Ran into the Nucleus

In all cases investigated, dissociation of RanGTP-transport factor complexes was accomplished by RanBP1 and RanGAP, indicating a general recycling mechanism (Bischoff and Görlich 1997; Deane et al. 1997; Görlich et al. 1997; Kutay et al. 1997; Paraskeva et al. 1999). Export of the RanGTP-transport factor complexes followed by dissociation in the cytoplasm would result in a depletion of RanGTP in the nucleus and an accumulation of RanGDP in the cytoplasm. To avoid breakdown of the RanGTP gradient essential for nuclear transport, a mechanism for efficient import of Ran into the nucleus must exist. The RanGDP-binding protein NTF2 (nuclear transport factor-2; Grundmann et al. 1988; Moore and Blobel 1994; Paschal and Gerace 1995) has been identified as the import factor for RanGDP (Ribbeck et al. 1998; Smith et al. 1998). Its three-dimensional structure is that of an αβ barrel that opens at one end to form a distinctive hydrophobic cavity (Bullock et al. 1996). Interaction with Ran involves mainly this cavity and its surrounding surface, and the switch II loop (residues 65–78) of Ran (Stewart et al. 1998; Kent et al. 1999). NTF2 mediates binding of RanGDP to the nuclear pore complex, and presumably subsequent translocation into the nucleus. This requires direct interaction of NTF2 and RanGDP, since mutated NTF2 incapable of binding to Ran does not support this import (Clarkson et al. 1997). At a separate site, NTF2 also interacts with nucleoporins p62 and Nsp1p (Clarkson et al. 1996). In the nucleus, GTP-dependent dissociation of the NTF2-RanGDP complex very likely is achieved by RanGEF-induced nucleotide exchange, since a defect of RanGEF results in an accumulation of Ran in the cytoplasm (Ren et al. 1993). RanGTP has no measurable affinity for NTF2 and is bound by importin-β-related transport factors favoring nuclear accumulation of Ran (Ribbeck et al. 1998; Smith et al. 1998).

11 Mog1 Induces Release of GTP from Ran

The nuclear GTP release factor Mog1 has not yet found its place in the nuclear transport scheme. It was identified in yeast as a suppressor of temperature-sensitive Ran mutants. (Oki et al. 1998). Deletion of *MOG1* causes temperature-sensitive growth and a defect in protein import, whereas the export of mRNA appears to be normal. Overexpression of NTF2 remedies the temperature-sensitive phenotype of the *mog1* deletion mutant. Mog1 protein from yeast or human (Oki and Nishimoto 2000; Steggerda and Paschal 2000) specifically binds to RanGTP and displaces the nucleotide. Excess free nucleotide has no effect on the complex, very much in contrast to the situation with RanGEF. The question arises, how such a nucleotide-free complex might be dissociated.

12 RanGTP in Mitosis

Recently it has been found that RanGTP and RanGEF are required for microtubule aster formation (Ohba et al. 1999), microtubule stability (Fleig et al. 2000), and formation of the mitotic spindle (Carazo-Salas et al. 1999; Wilde and Zheng 1999). Conversely, spindle assembly is dramatically disrupted when exogenous RanBP1 is added to mitotic *Xenopus* egg extracts (Kalab et al. 1999). In fission yeast, perturbations of the Ran GTPase system caused by mutation or overexpression of the RanBP1 homologue or several other regulatory proteins result in a unique terminal phenotype that includes condensed chromosomes and a fragmented nuclear envelope (Demeter et al. 1995). Formation of nuclear envelopes from *Xenopus* egg extracts requires cytosol and is inhibited by mutant forms of Ran that cannot bind or hydrolyze GTP, or by depletion of Ran or RanGEF from the assembly reaction (Zhang et al. 1999; Hetzer et al. 2000). Thus Ran-bound GTP and its hydrolysis appear to play a direct role in the regulation of mitosis, independent of nuclear transport, at a cell cycle stage when the nuclear membrane is disrupted. This may indicate a whole new set of functions for Ran. However, at face value the situation is paradoxical in the context of contemporary models for nucleocytoplasmic transport, in that RanBP1/RanBP2 together with RanGAP is thought to hydrolyze all cytoplasmic RanGTP. Additional regulatory factors would be required to prevent a mitotic short-circuit of the system.

References

Ahmadian MR, Stege P, Scheffzek K, Wittinghofer A (1997) Confirmation of the arginine-finger hypothesis for the GAP-stimulated GTP-hydrolysis reaction of Ras. Nat Struct Biol 4:686–689

Azuma Y, Seino H, Seki T, Uzawa S, Klebe C, Ohba T, Wittinghofer A, Hayashi N, Nishimoto T (1996) Conserved histidine residues of RCC1 are essential for nucleotide exchange on Ran. J Biochem Tokyo 120:82–91

Azuma Y, Renault L, Garcia-Ranea JA, Valencia A, Nishimoto T, Wittinghofer A (1999) Model of the Ran-RCC1 interaction using biochemical and docking experiments. J Mol Biol 289: 1119–1130

Bayer P, Arndt A, Metzger S, Mahajan R, Melchior F, Jaenicke R, Becker J (1998) Structure determination of the small ubiquitin-related modifier SUMO-1. J Mol Biol 280:275–286

Becker J, Melchior F, Gerke V, Bischoff FR, Ponstingl H, Wittinghofer A (1995) RNA1 encodes a GTPase-activating protein specific for Gsp1p, the Ran/TC4 homologue of *Saccharomyces cerevisiae*. J Biol Chem 270:11860–11865

Beddow AL, Richards SA, Orem NR, Macara IG (1995) The Ran/TC4 GTPase-binding domain: identification by expression cloning and caracterizartion of a conserved sequence motif. Proc Natl Acad Sci USA 92:3328–3332

Béraud-Dufour S, Robineau S, Chardin P, Paris S, Chabre M, Cherfils J, Antonny B (1998) A glutamic finger in the guanine nucleotide exchange factor ARNO displaces Mg^{2+} and the β-phosphate to destabilize GDP on ARF1. EMBO J 17:3651–3659

Bischoff FR, Görlich D (1997) RanBP1 is crucial for the release of RanGTP from importin beta-related nuclear transport factors. FEBS Lett 419:249–254

Bischoff FR, Ponstingl H (1991a) Catalysis of guanine nucleotide exchange on Ran by the mitotic regulator RCC1. Nature 354:80–82

Bischoff FR, Ponstingl H (1991b) Mitotic regulator protein RCC1 is complexed with a nuclear ras-related polypeptide. Proc Natl Acad Sci USA 88:10830–10834

Bischoff FR, Maier G, Tilz G, Ponstingl H (1990) A 47-kDa human nuclear protein recognized by antikinetochore autoimmune sera is homologous with the protein encoded by RCC1, a gene implicated in onset of chromosome condensation. Proc Natl Acad Sci USA 87:8617–8621

Bischoff FR, Klebe C, Kretschmer J, Wittinghofer A, Ponstingl H (1994) RanGAP1 induces GTPase activity of nuclear ras-related Ran. Proc Natl Acad Sci USA 91:2587–2591

Bischoff FR, Krebber H, Kempf T, Hermes I, Ponstingl H (1995a) Human RanGTPase activating protein RanGAP1 is a homologue of yeast Rna1p involved in mRNA processing and transport. Proc Natl Acad Sci USA 92:1749–1753

Bischoff FR, Krebber H, Smirnova E, Dong W, Ponstingl H (1995b) Co-activation of RanGTPase and inhibition of GTP dissociation by Ran.GTP binding protein RanBP1 EMBO J 14:705–715

Boguski MS, McCormick F (1993) Proteins regulating Ras and its relatives. Nature 366:643–654

Boriack-Sjodin PA, Margarit SM, Bar-Sagi D, Kuriyan J (1998) The structural basis of the activation of Ras by Sos. Nature 394:337–343

Bourne HR, Sanders DA, McCormick F (1990) The GTPase superfamily: a conserved switch for diverse cell functions. Nature 348:125–132

Bourne HR, Sanders DA, McCormick F (1991) The GTPase superfamily: conserved structure and molecular mechanism. Nature 349:117–127

Bressan A, Somma MP, Lewis J, Santolamazza C, Copeland NG, Gilbert DJ, Jenkins NA, Lavia P (1991) Characterization of the opposite-strand genes from the mouse bidirectionally transcribed HTF9 locus. Gene 103:201–209

Bullock TL, Clarkson WD, Kent HM, Stewart M (1996) The 1.6 angstrom resolution crystal structure of nuclear transport factor 2 (NTF2). J Mol Biol 260:422–431

Butler G, Wolfe KH (1994) Yeast homologue of mammalian Ran binding protein 1. BBA-Gene Struct Expr 1219:711–712

Carazo-Salas RE, Guarguaglini G, Gruss OJ, Segref A, Karsenti E, Mattaj IW (1999) Generation of GTP-bound Ran by RCC1 is required for chromatin-induced mitotic spindle formation. Nature 400:178–181

Chi NC, Adam EJH, Visser GD, Adam SA (1996) RanBP1 stabilizes the interaction of Ran with p97 in nuclear protein import. J Cell Biol 135:559–569

Chook YM, Blobel G (1999) Structure of the nuclear transport complex karyopherin-beta2-Ran.GppNHp. Nature 399:230–237

Cingolani G, Petosa C, Weis K, Muller CW (1999) Structure of importin-beta bound to the IBB domain of importin-alpha. Nature 399:221–229

Clarkson WD, Kent HM, Stewart M (1996) Separate binding sites on nuclear transport factor 2 (NTF2) for GDP-Ran and the phenylalanine-rich repeat regions of nucleoporins p62 and Nsp1p. J Mol Biol 263:517–524

Clarkson WD, Corbett AH, Paschal BM, Kent HM, McCoy AJ, Gerace L, Silver PA, Stewart M (1997) Nuclear protein import is decreased by engineered mutants of nuclear transport factor 2 (NTF2) that do not bind GDP-Ran. J Mol Biol 272:716–730

Corbett AH, Koepp DM, Schlenstedt G, Lee MS, Hopper AK, Silver PA (1995) Rna1p, a Ran/TC4 GTPase activating protein, is required for nuclear import. J Cell Biol 130:1017–1026

Coutavas E, Ren M, Oppenheim JD, D'Eustachio P, Rush MG (1993) Characterization of proteins that interact with the cell-cycle regulatory protein Ran/TC4. Nature 366:585–587

Deane R, Schäfer W, Zimmermann H-P, Mueller L, Görlich D, Prehn S, Ponstingl H, Bischoff FR (1997) Ran-binding protein 5 (RanBP5) is related to nuclear transport factor importin-β but interacts differently with RanBP1. Mol Cell Biol 17:5087–5096

Delphin C, Guan T, Melchior F, Gerace L (1997) RanGTP targets p97 to RanBP2, a filamentous protein localized at the cytoplasmic periphery of the nuclear pore complex. Mol Biol Cell 8:2379–2390

Demeter J, Morphew M, Sazer S (1995) A mutation in the RCC1-related protein pim1 results in nuclear envelope fragmentation in fission yeast. Proc Natl Acad Sci USA 92:1436–1440

Dickmanns A, Bischoff FR, Marshallsay C, Lührmann R, Ponstingl H, Fanning E (1996) The thermolability of nuclear protein import in tsBN2 cells is suppressed by microinjected Ran-GTP or Ran-GDP, but not by RanQ69L or RanT24 N. J Cell Sci 109:1449–1457

Drivas GT, Shih A, Coutavas E, Rush MG, D'Eustachio P (1990) Characterization of four novel ras-like genes expressed in a human teratocarcinoma cell line. Mol Cell Biol 10:1793–1798

Epps JL, Tanda S (1998) The *Drosophila semushi* mutation blocks nuclear import of Bicoid during embryogenesis. Curr Biol 8:1277–1280

Fleig U, Salus SS, Karig I, Sazer S (2000) The fission yeast Ran GTPase is required for microtubule integrity. J Cell Biol 151:1101–1112

Floer M, Blobel G (1996) The nuclear transport factor karyopherin β binds stoichiometrically to Ran-GTP and inhibits the Ran GTPase activating protein. J Biol Chem 271:5313–5316

Floer M, Blobel G, Rexach M (1997) Disassembly of RanGTP-karyopherin beta complex, an intermediate in nuclear protein import. J Biol Chem 272:19538–19546

Görlich D, Kutay U (1999) Transport between the cell nucleus and the cytoplasm. Annu Rev Cell Dev Biol 15:607–660

Görlich D, Vogel F, Mills AD, Hartmann E, Laskey RA (1995) Distinct functions for the two importin subunits in nuclear protein import. Nature 377:246–248

Görlich D, Panté N, Kutay U, Aebi U, Bischoff FR (1996) Identification of different roles for RanGDP and RanGTP in nuclear protein import. EMBO J 15:5584–5594

Görlich D, Dabrowski M, Bischoff FR, Kutay U, Bork P, Hartmann E, Prehn S, Izaurralde E (1997) A novel class of RanGTP binding proteins. J Cell Biol 138:65–80

Goldberg J (1998) Structural basis for activation of ARF GTPase: mechanisms of guanine nucleotide exchange and GTP-myristoyl switching. Cell 95:237–248

Grundmann U, Nerlich C, Rein T, Lottspeich F, Kupper HA (1988) Isolation of cDNA coding for the placental protein 15 (PP15). Nucleic Acids Res 16:4721

Haberland J, Gerke V (1999) Conserved charged residues in the leucine-rich repeat domain of the Ran GTPase activating protein are required for Ran binding and GTPase activation. Biochem J 343:653–662

Haberland J, Becker J, Gerke V (1997) The acidic C-terminal domain of Rna1p is required for the binding of RanGTP and for RanGAP activity. J Biol Chem 272:24717–2426

Hellmuth K, Lau D, Bischoff FR, Künzler M, Hurt E, Simos G (1998) Yeast Los1p has properties of an exportin-like nucleocytoplasmic transport factor for tRNA. Mol Cell Biol 18:6374–6386

Hetzer M, BilbaoCortes D, Walther TC, Gruss OJ, Mattaj IW (2000) GTP hydrolysis by Ran is required for nuclear envelope assembly. Mol Cell 5:1013–1024

Hieda M, Tachibana T, Yokoya F, Kose S, Imamoto N, Yoneda Y (1999) A monoclonal antibody to the COOH-terminal acidic portion of Ran inhibits both the recycling of Ran and nuclear protein import in living cells. J Cell Biol 144:645–655

Hillig RC, Renault L, Vetter IR, Drell T, Wittinghofer A, Becker J (1999) The crystal structure of rna1p: A new fold for a GTPase-activating protein. Mol Cell 3:781–791

Hopper AK, Traglia HM, Dunst RW (1990) The yeast RNA1 gene product necessary for RNA processing is located in the cytosol and apparently excluded from the nucleus. J Cell Biol 111:309–321

Izaurralde E, Kutay U, von Kobbe C, Mattaj IW, Görlich D (1997) The asymmetric distribution of the constituents of the Ran system is essential for transport into and out of the nucleus. EMBO J 16:6535–6547

Kadowaki T, Goldfarb D, Spitz LM, Tartakoff AM, Ohno M (1993) Regulation of RNA processing and transport by a nuclear guanine nucleotide release protein and members of the Ras superfamily. EMBO J 12:2929–2937

Kalab P, Pu RT, Dasso M (1999) The Ran GTPase regulates mitotic spindle assembly. Curr Biol 9:481–484

Kent HM, Moore MS, Quimby BB, Baker AME, McCoy AJ, Murphy GA, Corbett AH, Stewart M (1999) Engineered mutants in the switch II loop of Ran define the contribution made by key residues to the interaction with nuclear transport factor 2 (NTF2) and the role of this interaction in nuclear protein import. J Mol Biol 289:565–577

Klebe C, Bischoff FR, Ponstingl H, Wittinghofer A (1995a) Interaction of the nuclear GTP-binding protein Ran with its regulatory proteins RCC1 and RanGAP1. Biochemistry 34:639–647

Klebe C, Prinz H, Wittinghofer A, Goody RS (1995b) The kinetic mechanism of Ran-nucleotide exchange catalyzed by RCC1. Biochemistry 34:12543–12552

Künzler M, Gerstberger T, Stutz F, Bischoff FR, Hurt E (2000) Yeast Ran-binding protein 1 (Yrb1) shuttles between the nucleus and cytoplasm and is exported from the nucleus via a CRM1 (XPO1)-dependent pathway. Mol Cell Biol 20:4295–4308

Kuhlmann J, Macara I, Wittinghofer A (1997) Dynamic and equilibrium studies on the interaction of Ran with its effector RanBP1. Biochemistry 36:12027–12035

Kutay U, Bischoff FR, Kostka S, Kraft R, Görlich D (1997) Export of importin alpha from the nucleus is mediated by a specific nuclear transport factor. Cell 90:1061–1070

Loeb JD, Davis LI, Fink GR (1993) NUP2, a novel yeast nucleoporin, has functional overlap with other proteins of the nuclear pore complex. Mol Biol Cell 4:209–222

Lounsbury KM, Macara IG (1997) Ran-binding protein 1 (RanBP1) forms a ternary complex with Ran and karyopherin beta and reduces Ran GTPase-activating protein (RanGAP) inhibition by karyopherin beta. J Biol Chem 272:551–555

Lounsbury KM, Beddow AL, Macara IG (1994) A family of proteins that stabilize the Ran/TC4 GTPase in its GTP-bound conformation. J Biol Chem 269:11285–11290

Mahajan R, Delphin C, Guan T, Gerace L, Melchior F (1997) A small ubiquitin-related polypeptide involved in targeting RanGAP1 to nuclear pore complex protein RanBP2. Cell 88:97–107

Mattaj IW, Englmeier L (1998) Nucleocytoplasmic transport: The soluble phase. Annu Rev Biochem 67:265–306

Matunis MJ, Coutavas E, Blobel G (1996) A novel ubiquitin-like modification modulates the partitioning of the Ran-GTPase-activating protein RanGAP1 between the cytosol and the nuclear pore complex. J Cell Biol 135:1457–1470

Matunis MJ, Wu J, Blobel G (1998) SUMO-1 modification and its role in targeting the Ran GTPase-activating protein RanGAP1, to the nuclear pore complex. J Cell Biol 140:499–509

Maurer P, Redd M, Solsbacher J, Bischoff FR, Greiner M, Podtelejnikov AV, Matthias M, Stade K, Weis K, Schlenstedt G (2001) The nuclear export receptor Xpo1p forms distinct complexes with NES transport substrates and the yeast Ran binding protein 1 (Yrb1p). Mol Biol Cell 12:539–549

Melchior F, Weber K, Gerke V (1993) A functional homologue of the RNA1 gene product in Schizosaccharomyces pombe: purification, biochemical characterization, and identification of a leucine-rich repeat motif. Mol Biol Cell 4:569–581

Moore MS, Blobel G (1994) Purification of a Ran-interacting protein that is required for protein import into the nucleus. Proc Natl Acad Sci USA 91:10212–10216

Moroianu J, Hijikata M, Blobel G, Radu A (1995) Mammalian karyopherin alpha(1)beta and alpha(2)beta heterodimers: alpha(1) or alpha(2) subunit binds nuclear localization signal and beta subunit interacts with peptide repeat-containing nucleoporins. Proc Natl Acad Sci USA 92:6532–6536

Mossessova E, Gulbis JM, Goldberg J (1998) Structure of the guanine nucleotide exchange factor Sec7 domain of human Arno and analysis of the interaction with ARF GTPase. Cell 92:415–423

Mueller A, Cordes V, Bischoff FR, Ponstingl H (1998) Human RanBP3, a group of nuclear Ran-GTP binding proteins. FEBS Lett 427:330–336

Nishimoto T, Eilen E, Basilico C (1978) Premature chromosome condensation in a ts DNA- mutant of BHK cells. Cell 15:475–483

Noguchi E, Hayashi N, Nakashima N, Nishimoto T (1997) Yrb2p, a Nup2p-related yeast protein, has a functional overlap with Rna1p, a yeast Ran-GTPase-activating protein. Mol Cell Biol 17:2235–2246

Noguchi E, Saitoh YH, Sazer S, Nishimoto T (1999) Disruption of the YRB2 gene retards nuclear protein export, causing a profound mitotic delay, and can be rescued by overexpression of XPO1/CRM1. J Biochem Tokyo 125:574–585

Ohba T, Nakamura M, Nishitani H, Nishimoto T (1999) Self-organization of microtubule asters induced in *Xenopus* egg extracts by GTP-bound Ran. Science 284:1356–1358

Ohtsubo M, Okazaki H, Nishimoto T (1989) The RCC1 protein, a regulator for the onset of chromosome condensation locates in the nucleus and binds to DNA. J Cell Biol 109:1389–1397

Oki M, Nishimoto T (2000) Yrb1p interaction with the Gsp1p C terminus blocks Mog1p stimulation of GTP release from Gsp1p. J Biol Chem 275:32894–32900

Oki M, Noguchi E, Hayashi N, Nishimoto T (1998) Nuclear protein import, but not mRNA export, is defective in *all Saccharomyces cerevisiae* mutants that produce temperature-sensitive forms of the Ran GTPase homologue Gsp1p. Mol Gen Genet 257:624–634

Paraskeva E, Izaurralde E, Bischoff FR, Huber J, Kutay U, Hartmann E, Lührmann R, Görlich D (1999) CRM1-mediated recycling of snurportin 1 to the cytoplasm. J Cell Biol 145:255–264

Paschal BM, Gerace L (1995) Identification of NTF2, a cytosolic factor for nuclear import that interacts with nuclear pore complex protein p62. J Cell Biol 129:925–937

Plafker K, Macara IG (2000) Facilitated nucleocytoplasmic shuttling of the Ran binding protein RanBP1. Mol Cell Biol 20:3510–3521

Ren M, Drivas G, D'Eustachio P, Rush MG (1993) Ran/TC4: A small nuclear GTP-binding protein that regulates DNA synthesis. J Cell Biol 120:313–323

Renault L, Nassar N, Vetter I, Becker J, Klebe C, Roth M, Wittinghofer A (1998) The 1.7 Å crystal structure of the regulator of chromosome condensation (RCC1) reveals a seven-bladed propeller. Nature 392:97–101

Rexach M, Blobel G (1995) Protein import into nuclei: association and dissociation reactions involving transport substrate, transport factors, and nucleoporins. Cell 83:683–692

Ribbeck K, Lipowsky G, Kent HM, Stewart M, Gorlich D (1998) NTF2 mediates nuclear import of Ran. EMBO J 17:6587–6598

Richards SA, Lounsbury KM, Macara IG (1995) The C terminus of the nuclear RAN/TC4 GTPase stabilizes the GDP-bound state and mediates interactions with RCC1, RAN-GAP, and HTF9A/RANBP1. J Biol Chem 270:14405–14411

Scheffzek K, Klebe C, Fritz-Wolf K, Kabsch W, Wittinghofer A (1995) Crystal structure of the nuclear Ras-related protein Ran in its GDP-bound form. Nature 374:378–381

Scheffzek K, Lautwein A, Kabsch W, Ahmadian MR, Wittinghofer A (1996) Crystal structure of the GTPase-activating domain of human p120GAP and implications for the interaction with Ras. Nature 384:591–596

Scheffzek K, Ahmadian MR, Kabsch W, Wiesmuller L, Lautwein A, Schmitz F, Wittinghofer A (1997) The Ras-RasGAP complex: structural basis for GTPase activation and its loss in oncogenic Ras mutants. Science 277:333–338

Scheffzek K, Ahmadian MR, Wittinghofer A (1998) GTPase-activating proteins: helping hands to complement an active site. Trends Biochem Sci 23:257–262

Schlenstedt G, Wong DH, Koepp DM, Silver PA (1995) Mutants in a yeast Ran binding protein are defective in nuclear transport. EMBO J 14:5367–5378

Schlenstedt G, Smirnova E, Deane R, Solsbacher J, Kutay U, Görlich D, Ponstingl H, Bischoff FR (1997) Yrb4p, a yeast RanGTP-binding protein in import of ribosomal protein L25 into the nucleus. EMBO J 16:6237–6249

Singh BB, Patel HH, Roepman R, Schick D, Ferreira PA (1999) The zinc finger cluster domain of RanBP2 is a specific docking site for the nuclear export factor, exportin-1. J Biol Chem 274:37370–37378

Smith A, Brownawell A, Macara IG (1998) Nuclear import of Ran is mediated by the transport factor NTF2. Curr Biol 8:1403–1406

Steggerda SM, Paschal BM (2000) The mammalian Mog1 protein is a guanine nucleotide release factor for Ran. J Biol Chem 275:23175–23180

Stewart M, Kent HM, McCoy AJ (1998) Structural basis for molecular recognition between nuclear transport factor 2 (NTF2) and the GDP-bound form of the Ras-family GTPase Ran. J Mol Biol 277:635–646

Tachibana T, Imamoto N, Seino H, Nishimoto T, Yoneda Y (1994) Loss of RCC1 leads to suppression of nuclear protein import in living cells. J Biol Chem 269:24542–24545

Taura T, Schlenstedt G, Silver P (1997) Yrb2p is a nuclear protein that interacts with Prp20p, a yeast Rcc1 homologue. J Biol Chem 272:31877–31884

Taura T, Krebber H, Silver PA (1998) A member of the Ran-binding protein family, Yrb2p, is involved in nuclear protein export. Proc Natl Acad Sci USA 95:7427–7432

Turi TG, Mueller UW, Sazer S, Rose JK (1996) Characterization of a nuclear protein conferring brefeldin a resistance in *Schizosaccharomyces pombe*. J Biol Chem 271:9166–9171

Valencia A, Chardin P, Wittinghofer A, Sander C (1991) The ras protein family: evolutionary tree and role of conserved amino acids. Biochemistry 30:4637–4648

Vetter IR, Arndt A, Kutay U, Görlich D, Wittinghofer A (1999a) Structural view of the Ran-importin β interaction at 2.3 Å resolution. Cell 97:635–646

Vetter IR, Nowak C, Nishimoto T, Kuhlmann J, Wittinghofer A (1999b) Structure of a Ran-binding domain complexed with Ran bound to a GTP analogue: implications for nuclear transport. Nature 398:39–46

Villa Braslavsky CI, Nowak C, Görlich D, Wittinghofer A, Kuhlmann J (2000) Different structural and kinetic requirements for the interaction of Ran with the Ran-binding domains from RanBP2 and importin-beta. Biochemistry 39:11629–11639

Wilde A, Zheng Y (1999) Stimulation of microtubule aster formation and spindle assembly by the small GTPase Ran. Science 284:1359–1362

Wilken N, Senecal JL, Scheer U, Dabauvalle MC (1995) Localization of the Ran-GTP binding protein RanBP2 at the cytoplasmic side of the nuclear pore complex. Eur J Cell Biol 68:211–219

Wu J, Matunis MJ, Kraemer D, Blobel G, Coutavas E (1995) Nup358, a cytoplasmically exposed nucleoporin with peptide repeats, Ran-GTP binding sites, zinc fingers, a cyclophilin a homologous domain, and a leucine-rich region. J Biol Chem 270:14209–14213

Yokoyama N, Hayashi N, Seki T, Panté N, Ohba T, Nishii K, Kuma K, Hayashida T, Miyata T, Aebi U, Fukui M, Nishimoto T (1995) A giant nucleopore protein that binds Ran/TC4. Nature 376:184–188

Zhang CM, Hughes M, Clarke PR (1999) Ran-GTP stabilises microtubule asters and inhibits nuclear assembly in *Xenopus* egg extracts. J Cell Sci 112:2453–2461

Exportin-Mediated Nuclear Export of Proteins and Ribonucleoproteins

Maarten Fornerod[1] and Mutsuhito Ohno[1]

1 Introduction

Protein export from the nucleus of eukaryotic cells serves three main purposes: (1) to remove factors that are only transiently required in the nucleus – transcriptional signaling molecules fall into this category; (2) to participate in RNA export – all RNAs that leave the nucleus, do so in complex with proteins; (3) to recycle import factors that have been dissociated from their cargoes and have to return to the cytoplasm for a further round. A fourth conceivable function is to remove proteins that have unintentionally entered the nucleus, e.g. by diffusion or by inclusion into a reforming nucleus after mitosis.

Nuclear export of proteins is an active process, in the sense that it can occur against a concentration gradient. It requires a signal on the export substrate, which is recognized by a saturable soluble factor called an export receptor. The cargo–export receptor interaction may be bridged by a third component, defined as the export adaptor. Cargo-loaded export receptors pass the nuclear envelope through nuclear pore complexes (NPCs). This so-called translocation step must involve transient interactions with one or more nuclear pore components (nucleoporins). At the cytoplasmic side of the NPC, the export receptor/cargo complex is disassembled, and the receptor returns to the nucleus to assist in another export reaction.

There are several independent export pathways. Originally, these were identified by cross-competition studies between different types of RNA cargoes injected into *Xenopus* oocyte nuclei (Jarmolowski et al. 1994). Different pathways were found to exist for mRNA and tRNA, and U snRNAs. Independently, a viral mRNA export pathway was described that was dependent on the Rev protein of human immunodeficiency virus-1 (HIV-1) (reviewed in Cullen, this Vol.). In recent years, much progress has been made in identifying both signals and receptors for a number of these export pathways. Also, further independent export pathways have been characterized.

[1] EMBL Gene Expression Programme, Meyerhofstrasse 1, 69117 Heidelberg, Germany
Present address: M. Fornerod, Netherlands Cancer Institute – H4, Plesmanlaan 121, 1066 CX Amsterdam, The Netherlands
Present address: M. Ohno, Institute for Virus Research, Kyoto University, Kyoto 606, Japan

Results and Problems in Cell Differentiation, Vol. 35
K. Weis (Ed.): Nuclear Transport

In this review we will focus on export pathways that make use of one class of export receptors, the exportins, that are responsible for the majority of export pathways characterized to date. However, at least one important export pathway, that of mRNA (Izaurralde, this Vol.), appears in general not to be exportin-mediated.

2 Ran Is King

Exportins belong to a superfamily of transport receptors related to the import receptor importin-β (Fornerod et al. 1997b; Görlich et al. 1997). Importin-mediated nuclear import is discussed in an accompanying chapter (Jakel and Görlich, this Vol.). The distinguishing features of the importin-β family of transport receptors is their ability to bind to the small nuclear GTPase Ran in the GTP-bound form, to nucleoporins, and to their transport cargoes. Ran is discussed comprehensively in an accompanying review (Bischoff, this Vol.).

Ran is the master switch governing directionality of importin- and exportin-mediated transport. In summary, the GTP-bound form of Ran is predicted to be dominant in the nucleus, while the GDP-bound form is dominant in the cytoplasm. The nucleocytoplasmic RanGTP/RanGDP distribution follows from the localization of its cofactors. Ran's guanine nucleotide exchange factor (RanGEF) is nuclear, and, since cellular GTP is in excess over GDP, promotes exchange of RanGDP to RanGTP in the nucleus. Ran has very low intrinsic GTPase activity and Ran's GTPase activating protein (RanGAP), is cytoplasmic. This ensures that RanGTP is hydrolyzed in the cytoplasm to RanGDP. The accessory proteins Ran binding protein 1 (RanBP1) and 2 (RanBP2/Nup358) further increase the efficiency of RanGTP hydrolysis.

RanGTP imposes directionality to nuclear export since RanGTP/exportin/cargo heterotrimeric complexes are several orders of magnitude more stable than exportin/cargo heterodimers (Fig. 1). Due to the RanGTP gradient over the nuclear envelope, export complexes can be formed in the nucleus, but not in the cytoplasm (Fornerod et al. 1997a; Izaurralde et al. 1997; Kutay et al. 1997a; Richards et al. 1997). Conversely, RanGTP dissociates importin/cargo complexes, providing directionality to nuclear import as well (Jakel and Görlich, this volume). After translocation through the NPC, the trimeric export complex is destabilized by binding of RanBP1 or RanBP1-like domains in RanBP2/Nup358 (Floer and Blobel 1996; Bischoff and Görlich 1997; Floer et al. 1997; Kutay et al. 1997a, 1998; Askjaer et al. 1999). The export reaction is made irreversible by GTP hydrolysis of the RanBP1-bound RanGTP, stimulated by RanGAP1 (see Fig. 1).

3 Exportins

Five exportins have so far been identified: CRM1/exportin-1 (Fornerod et al. 1997a; Fukuda et al. 1997; Stade et al. 1997), CAS/Cse1 (Kutay et al. 1997a;

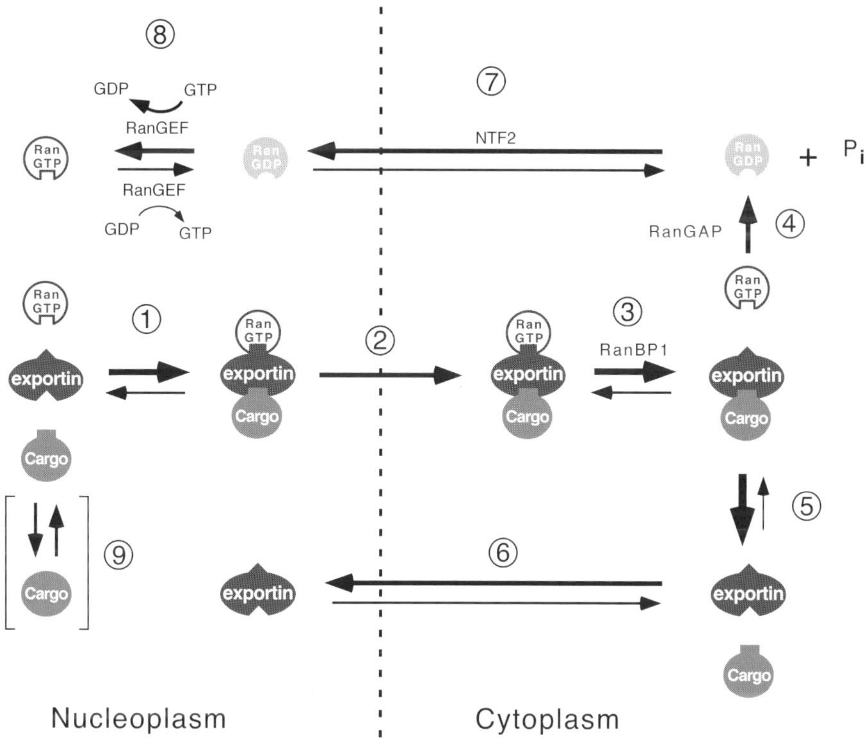

Fig. 1. Exportin-mediated nuclear export. A model for directionality of exportin-mediated nuclear export. *1* In the nucleus a trimeric complex forms between the exportin, nuclear export signal (NES)-cargo and RanGTP, promoted by high RanGTP concentration. *2* The trimeric complex traverses the nuclear pore complex (NPC), a process that is GTPase-independent. *3* In the cytoplasm RanBP1 or RanBP1-like domains in RanBP2/Nup358 bind to RanGTP and destabilize the trimeric complex. *4* Complex dissociation is made irreversible by RanGAP-stimulated GTP hydrolysis on Ran (see also Bischoff, this Vol.). *5* Exportin/cargo complexes have a low affinity in the absence of RanGTP. *6* The exportin recycles to the nucleus, and probably can shuttle without cargo. *7* RanGDP can diffuse through the NPC, but nuclear accumulation is stimulated by the import factor NTF2 (Görlich and Jakel, this Vol.). *8* In the nucleus, nucleotide exchange on Ran is mediated by RanGEF (RCC1 in higher eukaryotes). A higher GTP than GDP concentration in the cell ensures a preferential exchange to RanGTP. *9* Export signals may be regulated in the nucleoplasm. See text for references

Kunzler and Hurt 1998; Solsbacher et al. 1998), Los1/exportin-t (Arts et al. 1998; Hellmuth et al. 1998; Kutay et al. 1998), exportin-4 (Lipowsky et al. 2000) and Msn5 (Kaffman et al. 1998a), as summarized in Table 1. In the *Saccharomyces cerevisiae* proteome, 14 importin-β family members can be identified; nine have so far been characterized as import receptors (Jakel and Görlich, this Vol.), and four as export receptors (Table 1). For a detailed description of CAS and Los1p/exportin-t, we refer to Jakel and Görlich (this Vol.) and Hurt (this volume). In summary, CAS/Cse1 is the export receptor for the major NLS

Table 1. Exportins

Exportin[a]	Cargo
CRM1/exportin-1	NES proteins, U snRNAs, viral mRNAs, 5S rRNA, Snurportin1, RanBP1
CAS/Cse1	Importin α
Exportin-t/Los1	tRNAs
Exportin4	eIF5A
Msn5/exportin-5	Phosphorylated Pho4, Msn2, Msn4, Far1 and Mig1

[a] *S. cerevisiae* exportins are Crm1, Cse1, Los1 and Msn5. Crm1 is also called Xpo1; Msn5 is also known as Ste21. Metazoan exportins are CRM1/exportin-1, CAS, exportin-t and exportin-4, and a vertebrate homologue of Msn5. See text for details.

import adaptor importin-α/Srp1. The function of CAS/Cse1 is to return importin-α to the cytoplasm after NLS/importin-α/importin-β complex disassembly (Kutay et al. 1997a; Kunzler and Hurt 1998; Solsbacher et al. 1998). Vertebrate exportin-t and its *S. cerevisiae* homologue Los1p are export receptors for tRNAs (Arts et al. 1998; Hellmuth et al. 1998; Kutay et al. 1998). Both receptors only stably bind their export cargo in a trimeric complex with RanGTP (Kutay et al. 1997a, 1998; Arts et al. 1998).

4 CRM1/Exportin-1

CRM1 (for chromosome region maintenance) was originally identified in *Schizosaccharomyces pombe* by cold-sensitive mutations that resulted in premature chromosome condensation at the restrictive temperature (Adachi and Yanagida 1989). Independently, it was identified in a screen for genes that conferred resistance to the fungal cytotoxin leptomycin B (Nishi et al. 1994). The human homologue was found in complex with the nucleoporin CAN/Nup214 (Fornerod et al. 1997b), and was recognized to be distantly related to importin-β (Fornerod et al. 1997b; Görlich et al. 1997). When it was reported that leptomycin B could inhibit Rev-mediated RNA export but not mRNA export in cultured mammalian cells (Wolff et al. 1997), several laboratories found evidence that CRM1 was an export receptor for Rev-like leucine-rich export signals (Fornerod et al. 1997a; Fukuda et al. 1997; Ossareh-Nazari et al. 1997; Stade et al. 1997), demonstrating that nuclear export and import were related at both the structural and the functional level.

4.1 Leptomycin B and CRM1

The existence of a low molecular weight compound that specifically inhibits the CRM1 pathway has been of great value. Most eukaryotes are sensitive to

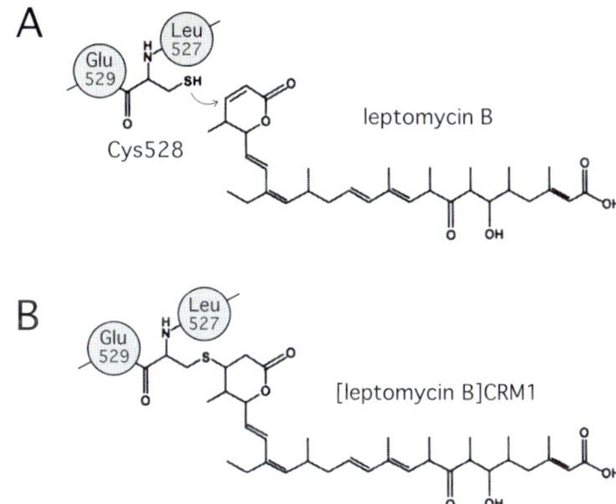

Fig. 2A, B. Leptomycin B is a specific inhibitor of CRM1. **A** Leptomycin B reacts with Cys-528 in human CRM1 (529 in *S. pombe*) through nucleophilic attack of the sulfhydryl side chain to an α,β-unsaturated carbonyl group of leptomycin B. Leptomycin B is proposed to initially bind to CRM1 via hydrophobic interactions of the branched fatty acid chain. **B** CRM1/leptomycin B reaction product. (Kudo et al. 1999a; Neville and Rosbash 1999)

leptomycin B at nanomolar concentrations, with the notable exception of *Saccharomyces cerevisiae*, which is completely resistant (Hamamoto et al. 1983a; Stade et al. 1997). Leptomycin B was isolated as an antifungal cytotoxin from a *Streptomyces* strain (Hamamoto et al. 1983a). It consists of an unsaturated, branched fatty acid chain with a terminal d-lactone ring (Fig. 2; Hamamoto et al. 1983b). It directly binds to CRM1, thereby interfering with formation of the RanGTP/CRM1/NES export complex (Fornerod et al. 1997b; Kudo et al. 1998). In fact, leptomycin B covalently attaches to a conserved cysteine corresponding to position 528 of the human protein and 529 of the *Schizosaccharomyces pombe* protein (Fig. 2; Kudo et al. 1999a; Neville and Rosbash 1999). The *S. cerevisiae* CRM1 has a threonine at the corresponding position (amino acid 539), explaining its leptomycin B resistance. Changing this threonine into a cysteine makes *S. cerevisiae* fully leptomycin B sensitive (Neville and Rosbash 1999). Conversely, mutation of *S. pombe* CRM1 Cys-529 into serine renders it leptomycin B resistant (Kudo et al. 1999a). CRM1 is the major protein that binds to biotinylated leptomycin B from HeLa cell extract (Kudo et al. 1998, 1999a). Together these data suggest very strongly that CRM1 is the only target for leptomycin B in the eukaryotic cell. For this reason leptomycin B has become a beloved tool to identify proteins that are exported via the CRM1 pathway, or to establish that a seemingly cytoplasmic protein in fact shuttles between the nucleus and the cytoplasm, sometimes pointing out unexpected nuclear functions. However, considering the many cellular pathways served by CRM1, it is not surprising that its inhibition can rapidly lead to indirect effects on other nuclear export events (see below).

4.2 Leucine-Rich Nuclear Export Signals

Prototype leucine-rich export signals that mediate rapid export from the nucleus were simultaneously discovered in the viral protein HIV-1 Rev and the cellular protein A phosphorylation inhibitor PKI (Fischer et al. 1995; Wen et al. 1995). In the case of Rev, the export signal had been previously characterized as a domain that was essential for the role of Rev in HIV-1 mRNA export (for a review of the role of Rev in HIV-1, see Cullen, this volume). The peptide signal was originally defined as four regularly spaced leucine residues following L-X_{2-3}-L-X_{2-3}-L-X-L, where L is leucine and X any amino acid. On the basis of this consensus, a number of putative NESs were identified in the proteome (Fritz and Green 1996). However, following functional characterization of these and other nuclear export signals, it has become clear that extreme care should be taken in assigning a nuclear export function to a leucine-rich sequence (see Table 2). Even when the leucine-rich peptide sequence can direct export of a reporter protein and is able to bind CRM1 in vitro (e.g. Ossareh-Nazari et al. 1997), this is not sufficient to conclude an export function in the context of the wild-type protein (Johnson et al. 1999; Huang et al. 2000). In fact, some proposed NESs were subsequently crystallized and shown to mediate intra-molecular hydrophobic surfaces (Iovine et al. 1997; Boche and Fanning 1997; Matunis et al. 1999). It is not unlikely that mutations in these surfaces cause a protein misfolding that indirectly causes an export defect. To complicate matters, sequence comparison of well-characterized functional NESs (Table 2) and random mutagenesis studies (Bogerd et al. 1996; Zhang and Dayton 1998) show that most leucines can be replaced by other hydrophobic amino acids (M, V, F, I, and occasionally W or C), except for the penultimate position that almost invariably is leucine or isoleucine. The intervening amino acids follow a certain loose pattern in that they are mostly charged, polar or small. Although the Rev NES functions in yeast (e.g. Stade et al. 1997), it is unknown to what extent yeast NESs conform to the higher eukaryote consensus (see Table 2). In vitro binding studies have shown that there are differences of over an order of magnitude in affinity to CRM1 among natural NES sequences (Askjaer et al. 1999; Henderson and Eleftheriou 2000). This suggests that the nuclear and cytoplasmic concentrations of shuttling proteins may be determined by the strength of their NESs.

4.3 Regulation of Leucine-Rich Nuclear Export Signals

An obvious way to regulate the nucleocytoplasmic distribution of a single shuttling protein is to regulate its NES. Indeed several key cell regulators have been proposed to make use of this mechanism, as detailed by Ruis and Schöller (this Vol.). For leucine-rich NESs, the *S. cerevisiae* protein yAP1 is a good example. This transcription factor activates several genes responsive to oxidative stress. It rapidly shuttles between the nucleus and cytoplasm due to an NLS and an

Table 2. Leucine-rich nuclear export signals

Protein[a]	Amino acid sequence[b]	Reference(s)
Viral		
HIV-1 Rev	L-PPL-ERLTL	Meyer and Malim (1994); Wen et al. (1995)
HTLV-I Rex	LSAQLYSSLSL	Kim et al. (1996); Palmeri and Malim (1996)
HSV-1 ICP27	L-IDLGLDLDL	Sandri-Goldin (1998)
MVM NS2	MTKKF-GTLTI	Askjaer et al. (1999); Ohshima et al. (1999)
Influenza NS1	F-DRL-ETLIL	Li et al. (1998)
Vertebrate		
PKI	LALKL-AGLDI	Wen et al. (1995)
MAPKK	LQKKL-EELEL	Fukuda et al. (1996)
c-Abl	LESNL-RELQI	Taagepera et al. (1998)
Cyclin B1	LCQAF-SDVIL	Toyoshima et al. (1998); Yang et al. (1998)
β-Actin NES1	LPHAI-MRLDL	Wada et al. (1998)
β-Actin NES2	IKEKL-CYVAL	Wada et al. (1998)
RanBP1[d]	VAEKL-EALSV	Richards et al. (1996); Zolotukhin and Felber (1997)
An3	LDQQF-AGLDL	Askjaer et al. (1999)
IκBα	MVKEL-QEIRL	Johnson et al. (1999); Huang et al. (2000)
p53	MFRELNEALEL	Stommel et al. (1999)
FMR	L-KEVDQLRL	Fridell et al. (1996)
HDM2	L-SFDESLAL	Roth et al. (1998)
Consensus[c]:	$\mathbf{\Phi}X_{2\text{-}3}\mathbf{\Phi}X_{2\text{-}3}\mathbf{\Phi}X\mathbf{\Phi}$	Bogerd et al. (1996); Kim et al. (1996); Zhang and Dayton (1998)
Yeast		
Yap1	IDVDGLCS	Yan et al. (1998)
Pap1	IDDLCSKLKN	Kudo et al. (1999b)
Non-consensus NESs[e]		
EIAV Rev	PLESDQWCRVLRQSL PEEKIP	Meyer et al. (1996); Harris et al. (1998)
FIV Rev	KKMMTDLEDRFRKLF GSPSKDEYT	Mancuso et al. (1994); Otero et al. (1998)
Ad E3–34kD	MVLTREELVI	Dobbelstein et al. (1997)
NFATc	SAIVAAINALTT	Klemm et al. (1997)

[a] A selection of viral, vertebrate and yeast proteins with CRM1-dependent export signals that have been confirmed within their natural context and fused to a heterologous protein. Viral proteins: *HIV-1* Human immunodeficiency virus 1, *HTLV-I* human T-cell lymphotropic virus type I, *HSV-1* herpes simplex virus 1, *MVM* minute virus of mice, *Ad* adenovirus type 5, *EIAV* equine infectious anemia virus, *FIV* feline immunodeficiency virus.

[b] NES sequence, mostly containing four characteristically spaced hydrophobic residues, indicated in bold. Leucines are most common, but other hydrophobic residues are also frequently found.

[c] Consensus based on these and phylogenetically close (data not shown) NESs of Rev-like spacing; see text for details.

[d] May not be a standard NES; see text for details.

[e] NESs that mediate CRM1-dependent export but do not conform to the classic leucine-rich NES consensus.

NES, and its export is mediated by Crm1p/Xpo1p (Kuge et al. 1998; Toone et al. 1998; Yan et al. 1998). Its NES contains several conserved cysteines, and binding to CRM1 in crude extracts is decreased under oxidative conditions (Yan et al. 1998). This suggests that oxidative stress decreases the strength of the yAP1 NES by oxidation of its cysteines, leading to the required increased nuclear concentration of the transcription factor. Phosphorylation of the stress response MAP kinase MK2 at a site near its proposed NES activates its nuclear export (Engel et al. 1998). Also PHAX, the export adaptor for U snRNA export (see below), needs to be phosphorylated in order to bind CRM1 (Ohno et al. 2000). Conversely, phosphorylation of cyclin B1 inhibits its export (Li et al. 1997). In these cases it has still to be determined whether the phosphorylation modifies the affinity of the NES for CRM1 per se, or influences its accessibility.

4.4 Three-Dimensional Structure of the Nuclear Export Signal

Perhaps due to the location of NESs in flexible regions, only four proposed NES structures have so far been solved: two in β-actin (Wada et al. 1998) and one each in p53 (Stommel et al. 1999) and in a 14–3-3 protein (Rittinger et al. 1999). The structure of these NESs are therefore not necessarily representative. Comparison of the NESs of the 14–3-3 protein and p53 indicated that hydrophobic side chains of the critical first three 3 residues form a hydrophobic stripe along an α-helix, while the fourth one faces almost opposite (Rittinger et al. 1999). However, the two NESs identified in β-actin do not seem to follow this topology at all (Wada et al. 1998). A fourth protein of known structure for which an NES has been postulated is the HIV-1 matrix protein (Dupont et al. 1999). Unfortunately, its NES does not seem to conform to the leucine-rich consensus sequence (Dupont et al. 1999). It will be of interest to see if there is indeed a fixed mode of binding of NESs to CRM1, or whether, like NLS importin-α interactions, different NESs exhibit unique interaction modes (see Conti, this volume).

4.5 Non-Consensus Nuclear Export Signals

The vertebrate U snRNP import adaptor Snurportin1 (Huber et al. 1998) is perhaps the best characterized example of a protein that is exported via the CRM1 pathway, but lacks a canonical NES (Paraskeva et al. 1999). It directly binds to CRM1 in a RanGTP-dependent way through a rather large domain, instead of a small peptide (Paraskeva et al. 1999). Binding is stronger than most NES substrates and competes very efficiently for U snRNA export (Paraskeva et al. 1999). Moreover, Rev-like proteins of the equine infectious anemia virus (EIAV) and feline immunodeficiency virus (FIV) contain a short NES that significantly differs from that of HIV-1 Rev and from cellular NESs in the spacing of their hydrophobic amino acids (Table 1). These NESs are functionally interchangeable with the HIV-1 Rev NES (Fridell et al. 1993; Mancuso et al. 1994) and mediate CRM1/exportin-1-dependent nuclear export (Otero et al. 1998). In these cases, direct binding to CRM1/exportin-1 has not been demonstrated. These three examples indicate that there is more than one way that a cargo can interact with CRM1 in a RanGTP-regulated manner. RNA aptamers have been described that specifically bind CRM1 and compete for NES binding (Hamm et al. 1997; Hamm and Fornerod 2000), raising the possibility that some RNAs may utilize directly the CRM1 pathway without a protein adaptor. Finally, it is good to keep in mind that alternative regions of CRM1 might be utilized by export cargoes to hitch a ride out of (or into) the nucleus. One problem may be that they could interfere with the standard cargoes or NPC binding. Another may be their inability to use the RanGTP switch for determining the direction of transport . Such "rogue" substrates may therefore have to use their own nucleocytoplasmic sensor to avoid being fruitlessly translocated back and forth.

4.6 Export of RanBP1 and RanGAP

Other atypical CRM1-mediated export events seem to govern export of two Ran cofactors: RanBP1/Yrb1, and RanGAP/Rna1. It is currently unexplained why these two proteins should enter the nucleus at all, since their cytoplasmic localization is thought to be essential for the RanGTP gradient across the nuclear envelope. One hypothesis is that they unintentionally diffuse in, and their export is merely nuclear clearance. However, nuclear import of RanBP1 is active (Plafker and Macara 2000), and RanGAP1 appears to have a non-classical nuclear localization signal (Matunis et al. 1998), suggesting that the two proteins may have a nuclear function. Evidence that RanBP1 is exported via CRM1 seems quite persuasive. First, an 11-amino-acid leucine-rich sequence in the C-terminus of RanBP1 conforms to the NES consensus (Table 2) and is necessary for cytoplasmic localization, although only a much larger fragment is sufficient (Richards et al. 1996; Zolotukhin and Felber 1997). Second, this 11-amino-acid domain can mediate nuclear export of a heterologous protein (Richards et al. 1996). Third, RanBP1 accumulates in the nucleus of mammalian cells upon leptomycin B treatment (Plafker and Macara 2000) and in the nucleus of *Xenopus* oocytes upon saturation of the CRM1 pathway by NES peptide conjugates (Pasquinelli et al. 1997). As discussed above, however, isolated NES-like peptides can easily function as NESs, even if they do not have this function in their natural context. Also, the C-terminal export domain of RanBP1 does not stimulate cooperative binding of RanGTP to CRM1 (Ulrike Kutay and Ralf Bischoff, personal communication), as would be expected from a normal NES protein (Askjaer et al. 1999; Paraskeva et al. 1999). Indeed, RanBP1 results in dissociation of RanGTP from CRM1, not its binding (Askjaer et al. 1999). The budding yeast homologue of RanBP1, Yrb1, lacks the

C-terminal export domain, yet accumulates in the nucleus in strains carrying *crm1* temperature-sensitive alleles at the restrictive temperature (Kunzler et al. 2000). Moreover, Yrb1 forms a RanGTP-dependent complex with CRM1 in vitro (Kunzler et al. 2000). Surprisingly, nuclear export of Yrb1 in a strain carrying a leptomycin B sensitive CRM1 allele is not inhibited by leptomycin B. In addition, RanGTP binding to the Ran binding domain of Yrb1 is necessary for CRM1 interaction (Kunzler et al. 2000). This suggests that the Yrb1/RanGTP interaction represents the RanGTP requirement for Yrb1 to bind to CRM1. Thus, Yrb1 may be exported by CRM1 in a novel fashion, independent of direct RanGTP-CRM1 binding. RanGAP is able to mediate dissociation of the trimeric complex (Kunzler et al. 2000), suggesting that its cytoplasmic localization ensures the transport direction of Yrb1/CRM1.

Less is known about the nuclear export of RanGAP1 and its yeast homologue Rna1p. Both proteins are cytoplasmic under normal conditions, although Rna1p can be detected in the nucleus when expressed in HeLa cells (Traglia et al. 1996). In budding yeast, Rna1p accumulates in the nucleus in a strain carrying the temperature-sensitive *crm1-1* allele at the restrictive temperature (Feng et al. 1999). Mutation of NES-like motifs has the same effect. However, a crystal structure of Rna1p indicates that the hydrophobic residues in these putative NESs, and also the ones predicted in mammalian RanGAP1 (Matunis et al. 1998), are part of hydrophobic core domains and their side chains are not exposed to the solvent (Hillig et al. 1999). It remains to be established which role CRM1 plays in RanGAP/Rna1p export and how NESs are recognized.

4.7 NES and RanGTP Interaction Domains in CRM1

We currently only have a very rough idea where in CRM1 the presumably hydrophobic contacts with the NES take place. Protein footprinting on CRM1 with Rev showed strong protection of endoprotease cleavage at Lys-810 and Asp-716 (Askjaer et al. 1998). A fragment encompassing amino acids 416–600 was able to bind specifically to NES-mimicking RNA aptamers (Hamm and Fornerod 2000). Similarly, a partially overlapping region of CRM1 (amino acids 566–720) was required to bind an NES-like peptide (Ossareh-Nazari and Dargemont 1999). Together, the NES binding region is most likely situated in the central most conserved region of CRM1. Interestingly, it includes the Cys-528 that is covalently modified by leptomycin B (Kudo et al. 1999a; Neville and Rosbash 1999), suggesting that this unique modification may directly interfere with substrate binding.

CRM1's RanGTP binding domain is most likely located at its N-terminal part, which has detectable, but often low, sequence homology with all importins and exportins (Fornerod et al. 1997b; Görlich et al. 1997). In importin-β, the region that is sufficient to bind RanGTP with high affinity is amino acid 1–364 (Kutay et al. 1997b; Vetter et al. 1999) and would correspond to approximately the first 470 amino acids in human CRM1. Indeed, amino acids 60–160 of human CRM1 were found to be required for RanGTP binding in vitro (Ossareh-Nazari and Dargemont 1999).

4.8 Additional Soluble Factors That May Be Required for NES Export

Although a skeletal mechanism for NES export that is backed by both functional and biochemical data seems to be in place (Fig. 1), other factors have been described that are likely to play additional roles. The best candidate is *S. cerevisiae* Yrb2, a nuclear protein that contains a RanBP1-like Ran binding domain (Noguchi et al. 1997; Taura et al. 1997). It selectively binds to CRM1 in vitro (Taura et al. 1998) and its deletion, although not lethal, causes an NES export defect that can be rescued by CRM1 over-expression (Taura et al. 1998; Noguchi et al. 1999). The vertebrate homologue, RanBP3, is also a nuclear protein (Mueller et al. 1998) and binds selectively to CRM1 (Ludwig Englmeier, M.F., Iain W. Mattaj, submitted). Another yeast gene implicated in the CRM1 pathway is *SAC3*, which is synthetically lethal with Yrb2 (Jones et al. 2000). Interestingly, its deletion, which, like Yrb2, does not prevent yeast growth, causes nuclear accumulation of some but not other NES proteins. Since Sac3 protein is localized at the nuclear rim, it may be a NPC component (Jones et al. 2000). Other nuclear pore components implicated in the CRM1 pathway will be discussed below.

One protein factor repeatedly proposed to be involved in NES-CRM1 mediated export is eIF5A (eukaryotic translation initiation factor 5A; Ruhl et al. 1993; Bevec et al. 1996; Rosorius et al. 1999). eIF5A had been reported to bind to the NES in Rev (Ruhl et al. 1993), but this has been contradicted by Henderson and Percipalle (1997) and more recently by Lipowsky et al. (2000). Nuclear export of eIF5A has been shown to be dependent on exportin-4 (see below), not CRM1 (Lipowsky et al. 2000). Together, this indicates that eIF5A has no direct effect on Rev export or on the CRM1 pathway. Lastly, a leptomycin B insensitive activity has been described in mammalian cytosol that stimulates export of an NES substrate in permeabilized cells and is distinct from CRM1 (Holaska and Paschal 1998). We must await further characterization of this activity to evaluate its significance for NES export in vivo.

5 Exportin-Mediated Export of Ribonucleoproteins

As far as we know, all natural RNAs are exported from the nucleus in the form of ribonucleoproteins (RNPs). Association of the RNA with proteins in the nucleus prior to export serves as a way to coordinate nuclear and cytoplasmic reactions and protects the RNA at the same time from undesired interactions. In addition, as seen below, it may provide another level of export regulation. So far, the exportins implicated in RNP export are exportin-t (Hurt, this Vol.) and CRM1.

5.1 Nuclear Export of U snRNAs

The RNP export pathway that is currently best characterized is that of U snRNAs, and therefore can be discussed in some detail. Major spliceosomal U snRNAs such as U1, U2, U4 and U5 are transcribed in the nucleus by RNA polymerase II and acquire an m^7G-cap structure. In Metazoa, this class of U snRNAs is initially exported from the nucleus. In the cytoplasm the RNAs form complexes with a group of proteins termed the Sm proteins and the cap is hypermethylated. The mature snRNPs are then imported back into the nucleus where they participate in pre-mRNA splicing reactions (Mattaj 1986; Lührmann et al. 1990).

It has been shown that the m^7G-cap structure of U snRNAs serves as an essential signal for their nuclear export. U snRNA export is inhibited if the RNA does not have the cap structure or if excess capped RNA competitors are microinjected into the nucleus (Hamm and Mattaj 1990; Jarmolowski et al. 1994). Export is dependent on interaction of the cap structure with the nuclear cap binding complex or CBC (Izaurralde et al. 1994, 1995). CBC is a heterodimeric complex composed of two subunits, CBP80 and CBP20, both of which are required for binding to the m^7G-cap structure (Izaurralde et al. 1994, 1995; Kataoka et al. 1995). Microinjection of antibodies against CBP20 that prevent CBC from binding to the cap inhibits U snRNA export specifically. These data indicate that CBC is an export factor for U snRNAs (Izaurralde et al. 1995).

U snRNA export is also specifically inhibited by inactivating CRM1 with leptomycin B, by saturating CRM1 with an excess of NES peptides (Fischer et al. 1995; Fornerod et al. 1997a) or by reducing the nuclear concentration of RanGTP (Izaurralde et al. 1997). Based on this information, the simplest model for U snRNA export would be that CBC bridges the interaction between U snRNA and CRM1, and binds cooperatively with RanGTP. However, CRM1 and CBC do not bind, and an additional factor is required for U snRNA export: a 55 kDa phosphoprotein termed PHAX (phosphorylated adaptor for RNA export; Ohno et al. 2000).

PHAX, CBC, CRM1, RanGTP and a capped RNA assemble into a large complex in vitro that may represent the U snRNA export complex (Fig. 3). Consistent with this, nuclear microinjection of PHAX protein leads to increased U snRNA export, while microinjection of anti-PHAX antibodies has the reverse effect. In both cases tRNA, mRNA or Rev export is largely unaffected. Every step in U snRNA export complex formation involves some degree of cooperativity. The consequence of this in vivo is a reduction of the export rate of PHAX and CBC in the absence of RNA substrates, thus reducing futile shuttling and competition between PHAX and other substrates for CRM1.

PHAX has to be phosphorylated to be able to interact with CRM1 (Fig. 3). PHAX contains a number of phosphorylation sites, the majority of which are serine residues, and an NES-like sequence that is essential for interaction with

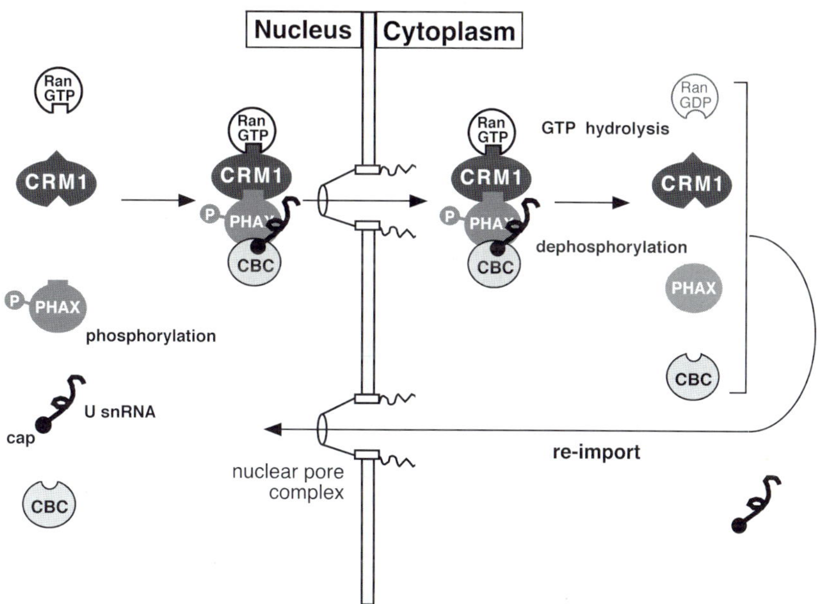

Fig. 3. A model of U snRNA export. An export complex containing CRM1, RanGTP, phosphorylated PHAX, CBC and U snRNA assembles in the nucleus. The complex subsequently moves through the NPC to the cytoplasmic side, where it disassembles due to GTP hydrolysis on Ran and dephosphorylation of PHAX. All the protein components then recycle back to the nucleus where PHAX is re-phosphorylated

CRM1 (Ohno et al. 2000). This putative NES, however, is some distance away from the major PHAX phosphorylation sites (M. Ohno, unpubl.). Phosphorylation of PHAX may change its conformation to expose the NES. Alternatively, the phosphorylated domain of PHAX in addition to the NES may contribute to the interaction with CRM1. PHAX phosphorylation takes place predominantly in the nucleus and dephosphorylation predominantly in the cytoplasm. Moreover, dephosphorylation of PHAX can trigger disassembly of the U snRNA export complex in vitro (Ohno et al. 2000). Therefore, the direction of U snRNA export can be ensured by a double mechanism: GTP hydrolysis on Ran and dephosphorylation of PHAX.

PHAX homologues are identifiable in the databases of many metazoan organisms but no obvious homologue is present in the *S. cerevisiae* or *S. pombe* genomes. U snRNP import in vertebrates involves importin-β (Palacios et al. 1997) and the U snRNP import adaptor Snurportin-1 (Huber et al. 1998). No obvious Snurportin-1 homologues are present in *S. cerevisiae* or *S. pombe* either. It therefore seems likely that U snRNP assembly may be a nuclear, not cytoplasmic process in at least these lower eukaryotes. It will be of interest to determine when and why the U snRNP maturation pathway changed.

5.2 Nuclear Export of 5S rRNA

In *Xenopus laevis*, large quantities of 5S ribosomal RNA are synthesized in excess over other ribosomal components and stored in the cytoplasm during early-stages oogenesis (Mairy and Denis 1972). Binding of newly synthesized 5S rRNA to either L5 or TFIIIA is necessary for nuclear export of the corresponding RNPs (Guddat et al. 1990). Export of 5S rRNA can be inhibited by saturation of the CRM1 pathway with NES conjugates (Fischer et al. 1995; Pasquinelli et al. 1997). TFIIIA contains a leucine-rich NES-like sequence that can replace the NES in HIV-1 Rev (Fridell et al. 1996). Together, this suggests that TFIIIA may function as a CRM1 export adaptor for 5S rRNPs. The mechanism of L5-mediated 5S rRNA export remains unclear.

5.3 Is CRM1 an Export Factor for mRNAs?

The short answer to this question is yes. HIV-1, HTLV-I and other complex retroviruses use Rev-like export adaptors for some of their mRNAs that feed into the CRM1 pathway (see also Cullen, this volume). Several other viruses also encode NES-containing proteins (Table 2), some of which may function in a similar way. Is CRM1 a general nuclear export factor for cellular mRNAs? Probably not, although the literature on this topic is contradictory. Saturation of the CRM1 pathway by NES peptide conjugates did (Pasquinelli et al. 1997) and did not (Fischer et al. 1995) inhibit mRNA export from *Xenopus* oocytes. Consistent with the latter result, leptomycin B or an excess of recombinant Snurportin1 did not inhibit export of the mRNA in *Xenopus* oocytes (Fornerod et al. 1997a; Paraskeva et al. 1999). Leptomycin B inhibition of CRM1 in mammalian cell culture does not yield a consensus either. Using a CAT reporter, Wolff (1997) found that leptomycin B caused a decrease in Rev-dependent but not in normal mRNA export. Watanabe et al. (1999), on the other hand, did detect both specific mRNA and poly(A) accumulation upon leptomycin B treatment. What shifts the balance in our opinion are observations made in *S. cerevisiae*. A yeast strain carrying the *crm1-1* temperature-sensitive mutation (Stade et al. 1997) or the leptomycin B sensitive *crm1* allele (Neville and Rosbash 1999) accumulates both NES proteins and poly(A) RNA in the nucleus. The very rapid inactivation of CRM1 by leptomycin B allows discrimination between these two effects in time. Whereas NES accumulation is already visible after 5 min, poly(A) accumulation only becomes apparent after 15–30 min. In comparison, the temperature-sensitive *rna1-1* mutation leads to a clear poly(A) RNA accumulation after only 2.5 min. Therefore it is likely that the poly(A) accumulation upon CRM1 inactivation is an indirect consequence of the NES export defect, and that the NES-CRM1 pathway is not a major export route in *S. cerevisiae*. These results make it also unlikely that a proposed NES in Gle1p (Murphy and Wente 1996) is essential for mRNA export. Interestingly, despite a strong poly(A) accumulation 6 h after leptomycin B addition, only a minor decrease in total protein synthesis was evident (Neville and Rosbash 1999). This indicates that poly(A) accumulation is not a good indicator for lack of mRNA export. Even though these experiments do not directly address the situation in metazoans, they suggest that the intermittently observed mRNA export defects upon CRM1 inactivation might also be indirect. Responsible for this may be the nuclear accumulation of RanGAP/Rna1p (Feng et al. 1999; Watanabe et al. 1999) or RanBP1/Yrb1p (Kunzler et al. 2000; Pasquinelli et al. 1997) upon CRM1 inactivation, since nuclear injection of these proteins in *Xenopus* oocytes interferes with mRNA export (Izaurralde et al. 1997). Also, nuclear accumulation of the mRNA export factor Dbp5p (Hodge et al. 1999) may be responsible. We refer to Izaurralde (this Vol.) for an extended discussion of the role of Dbp5 and the other factors implicated in general exportin-independent mRNA export. Nevertheless, it is very well possible that specific mRNAs are exported via the CRM1 pathway.

6 Exportin-4

In higher eukaryotes, exportin-4 mediates nuclear export of eIF5A(Lipowsky et al. 2000). In the presence of RanGTP, eIF5A was by far the most prominent protein selected from HeLa cell extracts on an exportin-4 column, while no import cargoes were selected in the absence of RanGTP. Moreover, exportin-4 stimulates eIF5A export in recombinant transport assays, whereas other exportins, including CRM1, do not. The export signal in eIF5A requires large parts of the protein and includes a hypusine modification that is unique to eIF5A (Lipowsky et al. 2000). It so far is the most distant member of the importin-β superfamily, and it is conserved among higher eukaryotes. Exportin-4 lacks a clear *S. cerevisiae* homologue, suggesting that eIF5A is not exported via this pathway in yeast, if at all. Even though eIF5A is one of the best conserved proteins in evolution (from bacteria to higher eukaryotes), its cellular role is largely unknown (Lipowsky et al. 2000 and references therein).

7 Msn5/Exportin-5

The fifth exportin is the *S. cerevisiae* Msn5 protein, also known as Ste21. It is not essential for growth but its deletion leads to defects in several signaling pathways (Alepuz et al. 1999). It has an as yet uncharacterized mammalian homologue, exportin-5. The model export cargo for Msn5 is Pho4, a transcription factor that is localized to the nucleus when cells are phosphate-starved, but translocates to the cytoplasm upon addition of phosphate to the medium, thereby turning off genes required for phosphate starvation (O'Neill et al. 1996; Kaffman et al. 1998a). Msn5 binds phosphorylated, but not unphosphorylated, Pho4 in a trimeric complex with RanGTP, and is essential for Pho4 nuclear exit (Kaffman et al. 1998a). This provides an explanation for its phosphate reactivity, because high phosphate medium activates the Pho80-Pho85 cyclin CDK complex, which phosphorylates Pho4 in the nucleus. Interestingly, Pho4 phosphorylation inhibits binding to its import receptor, Pse1 (Kaffman et al. 1998b), indicating a dual role for phosphorylation in promoting export and preventing reimport. A second cargo that has been identified is Far1p (Blondel et al. 1999). Also this cargo is translocated to the cytoplasm upon an extracellular signal, in this case the presence of pheromones. Other cargoes are Msn2, Msn4 and Mig1, all involved in signal transduction pathways. Although it is yet to be determined if there is a common signal between these different cargoes, it is conceivable that Msn5 is an export receptor dedicated to phosphorylated proteins.

8 Interaction of Exportins with the Nuclear Pore Complex

Relatively little mechanistic insight has been obtained on the translocation reaction of import and export receptors through the NPC. Functional studies in permeabilized HeLa cells have shown that CRM1-mediated NES export (and transportin-mediated M9 import) can occur in the absence of GTP hydrolysis on Ran or any other detectable form of energy (Englmeier et al. 1999; Ribbeck et al. 1999). Addition of high concentrations of a non-hydrolyzable form of RanGTP and NES cargo to permeabilized cells leads to reversal of export, i.e. CRM1-dependent NES import, revealing that there is no absolute directionality within the NPC (Nachury and Weis 1999). In *Xenopus* oocytes, leptomycin B has little or no influence on CRM1 shuttling (Fornerod et al. 1997a), indicating that association with an NES cargo or RanGTP is also not required for NPC translocation. Contacts between the NPC and CRM1 are likely to be mediated by phenylalanine-glycine (FG) repeats that are present in a subset of metazoan and yeast nucleoporins. These FG repeats are known to have a general affinity for importin-β-like transport receptors (see Fahrenkrog and Aebi, this volume). CRM1 interacts with the yeast nucleoporin RIP/Nup42 (Stutz et al. 1996), its vertebrate homologue NLP-1/CG1 (Farjot et al. 1999) and several other nucleoporin FG repeats (Neville et al. 1997; Floer and Blobel 1999; Zolotukhin and Felber 1999). The binding between CRM1 and the FG repeat of nucleoporin CAN/Nup214 is particularly strong (Fornerod et al. 1997b) and is increased further in the presence of RanGTP and NES cargo (Askjaer et al. 1999; Kehlenbach et al. 1999). CAN/Nup214 is located at the cytoplasmic face of the NPC on cytoplasmic filaments (see Fahrenkrog and Aebi, this Vol.). The strong RanGTP/CRM1/NES interaction with this nucleoporin may represent a termination site of NES export, since the Ran cofactors that are required for disassembly of the complex are nearby, attached to RanBP2/Nup358 (Fig. 4). RanBP2/Nup358 is localized on the cytoplasmic filaments, contains RanBP1-like domains (Wu et al. 1995; Yokoyama et al. 1995) and is bound to SUMO-modified RanGAP1 (Mahajan et al. 1997; Matunis et al. 1998; Saitoh et al. 1998). Additional RanGTP-dependent interactions with CRM1 have been detected

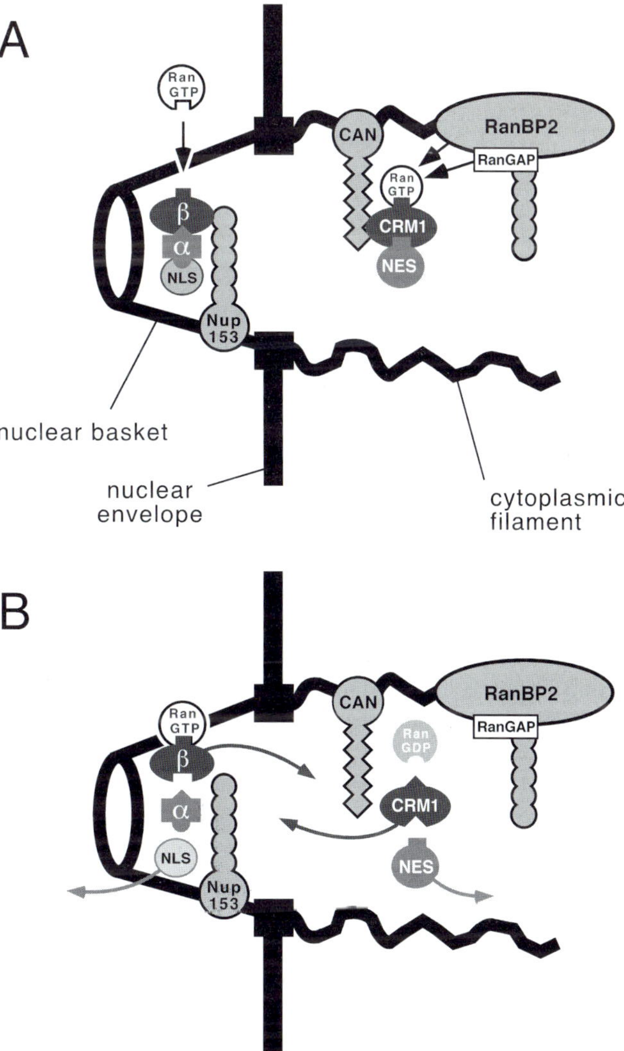

Fig. 4A, B. A model for disassembly of RanGTP/CRM1/NES export complexes at the cytoplasmic face of the NPC in higher eukaryotes. **A** After translocation, the trimeric export complex binds to the FG repeat region of nucleoporin CAN/Nup214. Disassembly is stimulated by RanBP1-like domains in nucleoporin RanBP2/Nup358, and SUMO-modified RanGAP1 (see also Bischoff, this Vol.), covalently bound to RanBP2/Nup358. The opposite situation may exist at the nuclear site of the NPC, where importin-β's stable association with Nup153 is disrupted by RanGTP (Shah et al. 1998). **B** Recycling of import and export receptor after release from the NPC, leaving their cargoes behind

with nucleoporins RIP/Nup42 (Floer and Blobel 1999), Nup153 (Nakielny et al. 1999) and Nup50/NPAP60 (Guan et al. 2000), suggesting that these nucleoporins are also involved in CRM1-mediated export. In fact, nuclear microinjection of antibodies to Nup50 inhibited CRM1-mediated NES export (Guan et al. 2000). However, mouse fibroblasts derived from Nup50 (–/–) embryos and lacking Nup50 were phenotypically normal and did not show a detectable export defect of the NES protein cyclin B1 (Smitherman et al. 2000). This suggests that the CRM1/Nup50 interaction is not essential for NES export, and more generally, that in vitro nucleoporin interactions and in vivo effects of nucleoporin antibody inhibition should be interpreted with care.

Interactions of the exportin Msn5 with the NPC were determined in vivo using fluorescence resonance energy transfer (FRET), and compared to the importin Pse1p (Damelin and Silver 2000). FRET signals are only detected if two proteins come into close proximity (1–10 nm), in this study defined as interaction. Of 13 nucleoporins tested, nine interacted with Msn5, and eight with Pse1. Interestingly, most of these were not FG-repeat-containing nucleoporins. Two interactions were specific for Msn5, with Nup82 and Nup84, whereas one was specific for Pse1, Nup53. Nup82 is in a subcomplex with the FG repeat nucleoporin Nup159 (Belgareh et al. 1998; Hurwitz et al. 1998), the closest yeast homologue of CAN/Nup214 (see Fahrenkrog and Aebi, this Vol.). It could therefore be speculated that the interaction with Nup159 is a terminal release step for the Msn5 export pathway, as has been suggested for the CAN/Nup214-CRM1 interaction (see above). It would be interesting to test the effect of the presence or absence of cargo on these FRET interactions. In conclusion, the interaction of exportins and importins with the NPC is still a rather blank spot in our understanding that is only slowly beginning to fill.

9 Perspectives

Much progress has been made since the identification of the first nuclear export pathways, but much still needs to be learned. It is clear that we have only started to understand the full spectrum of communication between the nucleus and the cytoplasm and the many pathways that are regulated in this way. As an increasing number of different importins and exportins are matched with their cargoes, it becomes clear that the interaction with the NPC remains a rich source of ignorance. One line of thought is that the NPC translocation involves "guided diffusion" through a wave of transient FG repeat–transport receptor interactions (Fahrenkrog and Aebi, this Vol.). If so, is there any specificity in these interactions, either between different transport receptors or between the same receptor loaded with different cargoes? And if there is specificity, is there regulation? What is the role of the stable NPC–transport receptor interactions? Answers to these questions must come from a combination of structural studies between NPC components and transport factors, biochemical binding studies using purified components, and biochemically manipulatable functional assays for NPC function. Another important question is whether there is regulation of transport at the level of the receptor? For CRM1 there is evidence that during certain stages of *Xenopus* embryonic development its export activity is down-regulated (Callanan et al. 2000). Other, more specialized transport receptors are likely to be more open to this type of regulation. It is our hope that by the continuing dissection of exportin-mediated export pathways and nucleocytoplasmic transport in general, we will gain a more complete understanding of the role of nucleocytoplasmic communication within the eukaryotic cell.

Acknowledgments. We thank Iain Mattaj, Kevin Czaplinski, Peter Askjaer, Scott Kuersten and Alexandra Segref for critically reading the manuscript and stimulating discussions. MF and MO were supported by a "Marie Curie" European Community post-doctoral fellowship and a grant from the Deutsches Forschungsgemeinschaft, respectively.

References

Adachi Y, Yanagida M (1989) Higher order chromosome structure is affected by cold-sensitive mutations in a *Schizosaccharomyces pombe* gene crm1+ which encodes a 115-kD protein preferentially localized in the nucleus and its periphery. J Cell Biol 108:1195–1207

Alepuz PM, Matheos D, Cunningham KW, Estruch F (1999) The *Saccharomyces cerevisiae* RanGTP-binding protein msn5p is involved in different signal transduction pathways. Genetics 153:1219–1231

Arts GJ, Fornerod M, Mattaj IW (1998) Identification of a nuclear export receptor for tRNA. Curr Biol 8:305–314

Askjaer P, Jensen TH, Nilsson J, Englmeier L, Kjems J (1998) The specificity of the CRM1-Rev nuclear export signal interaction is mediated by RanGTP. J Biol Chem 273:33414–33422

Askjaer P, Bachi A, Wilm M, Bischoff FR, Weeks DL, Ogniewski V, Ohno M, Niehrs C, Kjems J, Mattaj IW, Fornerod M (1999) RanGTP-regulated interactions of CRM1 with nucleoporins and a shuttling DEAD-box helicase. Mol Cell Biol 19:6276–6285

Belgareh N, Snay-Hodge C, Pasteau F, Dagher S, Cole CN, Doye V (1998) Functional characterization of a Nup159p-containing nuclear pore subcomplex. Mol Biol Cell 9:3475–3492

Bevec D, Jaksche H, Oft M, Wohl T, Himmelspach M, Pacher A, Schebesta M, Koettnitz K, Dobrovnik M, Csonga R, Lottspeich F, Hauber J (1996) Inhibition of HIV-1 replication in lymphocytes by mutants of the Rev cofactor eIF-5 A. Science 271:1858–1860

Bischoff FR, Görlich D (1997) RanBP1 is crucial for the release of RanGTP from importin beta-related nuclear transport factors. FEBS Lett 419:249–254

Blondel M, Alepuz PM, Huang LS, Shaham S, Ammerer G, Peter M (1999) Nuclear export of Far1p in response to pheromones requires the export receptor Msn5p/Ste21p. Genes Dev 13: 2284–2300

Boche I, Fanning E (1997) Nucleocytoplasmic recycling of the nuclear localization signal receptor alpha subunit in vivo is dependent on a nuclear export signal, energy, and RCC1. J Cell Biol 139:313–325

Bogerd HP, Fridell RA, Benson RE, Hua J, Cullen BR (1996) Protein sequence requirements for function of the human T-cell leukemia virus type 1 Rex nuclear export signal delineated by a novel in vivo randomization-selection assay. Mol Cell Biol 16:4207–4214

Callanan M, Kudo N, Gout S, Brocard M, Yoshida M, Dimitrov S, Khochbin S (2000) Developmentally regulated activity of CRM1/XPO1 during early *Xenopus* embryogenesis. J Cell Sci 113:451–459

Damelin M, Silver PA (2000) Mapping interactions between nuclear transport factors in living cells reveals pathways through the nuclear pore complex. Mol Cell 5:133–140

Dobbelstein M, Roth J, Kimberly WT, Levine AJ, Shenk T (1997) Nuclear export of the E1B 55-kDa and E4 34-kDa adenoviral oncoproteins mediated by a rev-like signal sequence. EMBO J 16:4276–4284

Dupont S, Sharova N, DeHoratius C, Virbasius CM, Zhu X, Bukrinskaya AG, Stevenson M, Green MR (1999) A novel nuclear export activity in HIV-1 matrix protein required for viral replication. Nature 402:681–685

Engel K, Kotlyarov A, Gaestel M (1998) Leptomycin B-sensitive nuclear export of MAPKAP kinase 2 is regulated by phosphorylation. EMBO J 17:3363–3371

Englmeier L, Olivo JC, Mattaj IW (1999) Receptor-mediated substrate translocation through the nuclear pore complex without nucleotide triphosphate hydrolysis. Curr Biol 9:30–41

Farjot G, Sergeant A, Mikaelian I (1999) A new nucleoporin-like protein interacts with both HIV-1 Rev nuclear export signal and CRM-1. J Biol Chem 274:17309–17317

Feng W, Benko AL, Lee JH, Stanford DR, Hopper AK (1999) Antagonistic effects of NES and NLS motifs determine *S. cerevisiae* Rna1p subcellular distribution. J Cell Sci 112:339–347

Fischer U, Huber J, Boelens WC, Mattaj IW, Lührmann R (1995) The HIV-1 Rev activation domain is a nuclear export signal that accesses an export pathway used by specific cellular RNAs. Cell 82:475–483

Floer M, Blobel G (1996) The nuclear transport factor karyopherin beta binds stoichiometrically to Ran-GTP and inhibits the Ran GTPase activating protein. J Biol Chem 271:5313–5316

Floer M, Blobel G (1999) Putative reaction intermediates in Crm1-mediated nuclear protein export. J Biol Chem 274:16279–16286

Floer M, Blobel G, Rexach M (1997) Disassembly of RanGTP-karyopherin beta complex, an intermediate in nuclear protein import. J Biol Chem 272:19538–19546

Fornerod M, Ohno M, Yoshida M, Mattaj IW (1997a) CRM1 is an export receptor for leucine-rich nuclear export signals. Cell 90:1051–1060

Fornerod M, van Deursen J, van Baal S, Reynolds A, Davis D, Murti KG, Fransen J, Grosveld G (1997b) The human homologue of yeast CRM1 is in a dynamic subcomplex with CAN/Nup214 and a novel nuclear pore component Nup88. EMBO J 16:807–816

Fridell RA, Partin KM, Carpenter S, Cullen BR (1993) Identification of the activation domain of equine infectious anemia virus rev. J Virol 67:7317–7323

Fridell RA, Fischer U, Lührmann R, Meyer BE, Meinkoth JL, Malim MH, Cullen BR (1996) Amphibian transcription factor IIIA proteins contain a sequence element functionally equivalent to the nuclear export signal of human immunodeficiency virus type 1 Rev. Proc Natl Acad Sci USA 93:2936–2940

Fritz CC, Green MR (1996) HIV Rev uses a conserved cellular protein export pathway for the nucleocytoplasmic transport of viral RNAs. Curr Biol 6:848–854

Fukuda M, Gotoh I, Gotoh Y, Nishida E (1996) Cytoplasmic localization of mitogen-activated protein kinase kinase directed by its NH2-terminal, leucine-rich short amino acid sequence, which acts as a nuclear export signal. J Biol Chem 271:20024–20028

Fukuda M, Asano S, Nakamura T, Adachi M, Yoshida M, Yanagida M, Nishida E (1997) CRM1 is responsible for intracellular transport mediated by the nuclear export signal. Nature 390:308–311

Görlich D, Dabrowski M, Bischoff FR, Kutay U, Bork P, Hartmann E, Prehn S, Izaurralde E (1997) A novel class of RanGTP binding proteins. J Cell Biol 138:65–80

Guan T, Kehlenbach RH, Scirmer EC, Kehlenbach A, Fan F, Clurman BE, Arnheim N, Gerace L (2000) Nup50, a nucleoplasmically oriented nucleoporin with a role in nuclear export. Mol Cell Biol 20:5619–5630

Guddat U, Bakken AH, Pieler T (1990) Protein-mediated nuclear export of RNA: 5S rRNA containing small RNPs in *Xenopus* oocytes. Cell 60:619–628

Hamamoto T, Gunji S, Tsuji H, Beppu T (1983a) Leptomycins A and B, new antifungal antibiotics. I. Taxonomy of the producing strain and their fermentation, purification and characterization. J Antibiot (Tokyo) 36:639–645

Hamamoto T, Seto H, Beppu T (1983b) Leptomycins A and B, new antifungal antibiotics. II. Structure elucidation. J Antibiot (Tokyo) 36:646–650

Hamm J, Fornerod M (2000) Anti-idiotype RNAs that mimic the leucine-rich nuclear export signal and specifically bind to CRM1/exportin 1. Chem Biol 7:345–354

Hamm J, Mattaj IW (1990) Monomethylated cap structures facilitate RNA export from the nucleus. Cell 63:109–118

Hamm J, Huber J, Lührmann R (1997) Anti-idiotype RNA selected with an anti-nuclear export signal antibody is actively transported in oocytes and inhibits Rev- and cap-dependent RNA export. Proc Natl Acad Sci USA 94:12839–12844

Harris ME, Gontarek RR, Derse D, Hope TJ (1998) Differential requirements for alternative splicing and nuclear export functions of equine infectious anemia virus Rev protein. Mol Cell Biol 18:3889–3899

Hellmuth K, Lau DM, Bischoff FR, Kunzler M, Hurt E, Simos G (1998) Yeast los1p has properties of an exportin-like nucleocytoplasmic transport factor for tRNA. Mol Cell Biol 18:6374–6386

Henderson BR, Eleftheriou A (2000) A comparison of the activity, sequence specificity, and CRM1-dependence of different nuclear export signals. Exp Cell Res 256:213–224

Henderson BR, Percipalle P (1997) Interactions between HIV Rev and nuclear import and export factors: the Rev nuclear localisation signal mediates specific binding to human importin-beta. J Mol Biol 274:693–707

Hillig RC, Renault L, Vetter IR, Drell Tt, Wittinghofer A, Becker J (1999) The crystal structure of rna1p: a new fold for a GTPase-activating protein. Mol Cell 3:781–791

Hodge CA, Colot HV, Stafford P, Cole CN (1999) Rat8p/Dbp5p is a shuttling transport factor that interacts with Rat7p/Nup159p and Gle1p and suppresses the mRNA export defect of xpo1-1 cells. EMBO J 18:5778–5788

Holaska JM, Paschal BM (1998) A cytosolic activity distinct from crm1 mediates nuclear export of protein kinase inhibitor in permeabilized cells. Proc Natl Acad Sci USA 95:14739–14744

Huang TT, Kudo N, Yoshida M, Miyamoto S (2000) A nuclear export signal in the N-terminal regulatory domain of IkappaBalpha controls cytoplasmic localization of inactive NF-kappaB/IkappaBalpha complexes. Proc Natl Acad Sci USA 97:1014–1019

Huber J, Cronshagen U, Kadokura M, Marshallsay C, Wada T, Sekine M, Lührmann R (1998) Snurportin1, an m3G-cap-specific nuclear import receptor with a novel domain structure. EMBO J 17:4114–4126

Hurwitz ME, Strambio-de-Castillia C, Blobel G (1998) Two yeast nuclear pore complex proteins involved in mRNA export form a cytoplasmically oriented subcomplex. Proc Natl Acad Sci USA 95:11241–11245

Iovine MK, Wente SR (1997) A nuclear export signal in Kap95p is required for both recycling the import factor and interaction with the nucleoporin GLFG repeat regions of Nup116p and Nup100p. J Cell Biol 137:797–811

Izaurralde E, Lewis J, McGuigan C, Jankowska M, Darzynkiewicz E, Mattaj IW (1994) A nuclear cap binding protein complex involved in pre-mRNA splicing. Cell 78:657–668

Izaurralde E, Lewis J, Gamberi C, Jarmolowski A, McGuigan C, Mattaj IW (1995) A cap-binding protein complex mediating U snRNA export. Nature 376:709–712

Izaurralde E, Kutay U, von Kobbe C, Mattaj IW, Görlich D (1997) The asymmetric distribution of the constituents of the Ran system is essential for transport into and out of the nucleus. EMBO J 16:6535–6547

Jarmolowski A, Boelens WC, Izaurralde E, Mattaj IW (1994) Nuclear export of different classes of RNA is mediated by specific factors. J Cell Biol 124:627–635

Johnson C, Van Antwerp D, Hope TJ (1999) An N-terminal nuclear export signal is required for the nucleocytoplasmic shuttling of IkappaBalpha. EMBO J 18:6682–6693

Jones AL, Quimby BB, Hood JK, Ferrigno P, Keshava PH, Silver PA, Corbett AH (2000) SAC3 may link nuclear protein export to cell cycle progression. Proc Natl Acad Sci USA 97:3224–3229

Kaffman A, Rank NM, O'Neill EM, Huang LS, O'Shea EK (1998a) The receptor Msn5 exports the phosphorylated transcription factor Pho4 out of the nucleus. Nature 396:482–486

Kaffman A, Rank NM, O'Shea EK (1998b) Phosphorylation regulates association of the transcription factor Pho4 with its import receptor Pse1/Kap121. Genes Dev 12:2673–2683

Kataoka N, Ohno M, Moda I, Shimura Y (1995) Identification of the factors that interact with NCBP, an 80 kDa nuclear cap binding protein. Nucleic Acids Res 23:3638–3641

Kehlenbach RH, Dickmanns A, Kehlenbach A, Guan T, Gerace L (1999) A role for RanBP1 in the release of CRM1 from the nuclear pore complex in a terminal step of nuclear export. J Cell Biol 145:645–657

Kim FJ, Beeche AA, Hunter JJ, Chin DJ, Hope TJ (1996) Characterization of the nuclear export signal of human T-cell lymphotropic virus type 1 Rex reveals that nuclear export is mediated by position-variable hydrophobic interactions. Mol Cell Biol 16:5147–5155

Klemm JD, Beals CR, Crabtree GR (1997) Rapid targeting of nuclear proteins to the cytoplasm. Curr Biol 7:638–644

Kudo N, Wolff B, Sekimoto T, Schreiner EP, Yoneda Y, Yanagida M, Horinouchi S, Yoshida M (1998) Leptomycin B inhibition of signal-mediated nuclear export by direct binding to CRM1. Exp Cell Res 242:540–547

Kudo N, Matsumori N, Taoka H, Fujiwara D, Schreiner EP, Wolff B, Yoshida M, Horinouchi S (1999a) Leptomycin B inactivates CRM1/exportin 1 by covalent modification at a cysteine residue in the central conserved region. Proc Natl Acad Sci USA 96:9112–9117

Kudo N, Taoka H, Toda T, Yoshida M, Horinouchi S (1999b) A novel nuclear export signal sensitive to oxidative stress in the fission yeast transcription factor Pap1. J Biol Chem 274: 15151–15158

Kuge S, Toda T, Iizuka N, Nomoto A (1998) Crm1 (XpoI) dependent nuclear export of the budding yeast transcription factor yAP-1 is sensitive to oxidative stress. Genes Cells 3:521–532

Kunzler M, Hurt EC (1998) Cse1p functions as the nuclear export receptor for importin alpha in yeast. FEBS Lett 433:185–190

Kunzler M, Gerstberger T, Stutz F, Bischoff FR, Hurt E (2000) Yeast ran-binding protein 1 (Yrb1) shuttles between the nucleus and cytoplasm and is exported from the nucleus via a CRM1 (XPO1)-dependent pathway. Mol Cell Biol 20:4295–4308

Kutay U, Bischoff FR, Kostka S, Kraft R, Görlich D (1997a) Export of importin alpha from the nucleus is mediated by a specific nuclear transport factor. Cell 90:1061–1071

Kutay U, Izaurralde E, Bischoff FR, Mattaj IW, Görlich D (1997b) Dominant-negative mutants of importin-beta block multiple pathways of import and export through the nuclear pore complex. EMBO J 16:1153–1163

Kutay U, Lipowsky G, Izaurralde E, Bischoff FR, Schwarzmaier P, Hartmann E, Görlich D (1998) Identification of a tRNA-specific nuclear export receptor. Mol Cell 1:359–369

Li J, Meyer AN, Donoghue DJ (1997) Nuclear localization of cyclin B1 mediates its biological activity and is regulated by phosphorylation. Proc Natl Acad Sci USA 94:502–507

Li Y, Yamakita Y, Krug RM (1998) Regulation of a nuclear export signal by an adjacent inhibitory sequence: the effector domain of the influenza virus NS1 protein. Proc Natl Acad Sci USA 95:4864–4869

Lipowsky G, Bischoff RF, Schwartzmaier P, Kraft R, Kostka S, Hartmann E, Kutay U, Görlich D (2000) Exportin 4: a mediator of a novel nuclear export pathway in higher eukaryotes. EMBO J (in press)

Lührmann R, Kastner B, Bach M (1990) Structure of spliceosomal snRNPs and their role in pre-mRNA splicing. Biochim Biophys Acta 1087:265–292

Mahajan R, Delphin C, Guan T, Gerace L, Melchior F (1997) A small ubiquitin-related polypeptide involved in targeting RanGAP1 to nuclear pore complex protein RanBP2. Cell 88:97–107

Mairy M, Denis H (1972) Biochemical studies on oogenesis. 2. Ribosome assembly during the development of oocytes in *Xenopus laevis*. Eur J Biochem 25:535–543

Mancuso VA, Hope TJ, Zhu L, Derse D, Phillips T, Parslow TG (1994) Posttranscriptional effector domains in the Rev proteins of feline immunodeficiency virus and equine infectious anemia virus. J Virol 68:1998–2001

Mattaj IW (1986) Cap trimethylation of U snRNA is cytoplasmic and dependent on U snRNP protein binding. Cell 46:905–911

Matunis MJ, Wu J, Blobel G (1998) SUMO-1 modification and its role in targeting the Ran GTPase-activating protein, RanGAP1, to the nuclear pore complex. J Cell Biol 140:499–509

Meyer BE, Malim MH (1994) The HIV-1 Rev *trans*-activator shuttles between the nucleus and the cytoplasm. Genes Dev 8:1538–1547

Meyer BE, Meinkoth JL, Malim MH (1996) Nuclear transport of human immunodeficiency virus type 1, visna virus, and equine infectious anemia virus Rev proteins: identification of a family of transferable nuclear export signals. J Virol 70:2350–2359

Mueller L, Cordes VC, Bischoff FR, Ponstingl H (1998) Human RanBP3, a group of nuclear RanGTP binding proteins. FEBS Lett 427:330–336

Murphy R, Wente SR (1996) An RNA-export mediator with an essential nuclear export signal. Nature 383:357–360

Nachury MV, Weis K (1999) The direction of transport through the nuclear pore can be inverted. Proc Natl Acad Sci USA 96:9622–9627

Nakielny S, Shaikh S, Burke B, Dreyfuss G (1999) Nup153 is an M9-containing mobile nucleoporin with a novel Ran-binding domain. EMBO J 18:1982–1995

Neville M, Rosbash M (1999) The NES-Crm1p export pathway is not a major mRNA export route in *Saccharomyces cerevisiae*. EMBO J 18:3746–3756

Neville M, Stutz F, Lee L, Davis LI, Rosbash M (1997) The importin-beta family member Crm1p bridges the interaction between Rev and the nuclear pore complex during nuclear export. Curr Biol 7:767–775

Nishi K, Yoshida M, Fujiwara D, Nishikawa M, Horinouchi S, Beppu T (1994) Leptomycin B targets a regulatory cascade of crm1, a fission yeast nuclear protein, involved in control of higher order chromosome structure and gene expression. J Biol Chem 269:6320–6324

Noguchi E, Hayashi N, Nakashima N, Nishimoto T (1997) Yrb2p, a Nup2p-related yeast protein, has a functional overlap with Rna1p, a yeast Ran-GTPase-activating protein. Mol Cell Biol 17:2235–2246

Noguchi E, Saitoh Y, Sazer S, Nishimoto T (1999) Disruption of the YRB2 gene retards nuclear protein export, causing a profound mitotic delay, and can be rescued by overexpression of XPO1/CRM1. J Biochem (Tokyo) 125:574–585

Ohno M, Segref A, Bachi A, Wilm M, Mattaj IW (2000) PHAX, a mediator of U snRNA nuclear export whose activity is regulated by phosphorylation. Cell 101:187–198

Ohshima T, Nakajima T, Oishi T, Imamoto N, Yoneda Y, Fukamizu A, Yagami K (1999) CRM1 mediates nuclear export of nonstructural protein 2 from parvovirus minute virus of mice. Biochem Biophys Res Commun 264:144–150

O'Neill EM, Kaffman A, Jolly ER, O'Shea EK (1996) Regulation of PHO4 nuclear localization by the PHO80-PHO85 cyclin-CDK complex. Science 271:209–212

Ossareh-Nazari B, Dargemont C (1999) Domains of Crm1 involved in the formation of the Crm1, RanGTP, and leucine-rich nuclear export sequences trimeric complex. Exp Cell Res 252: 236–241

Ossareh-Nazari B, Bachelerie F, Dargemont C (1997) Evidence for a role of CRM1 in signal-mediated nuclear protein export. Science 278:141–144

Otero GC, Harris ME, Donello JE, Hope TJ (1998) Leptomycin B inhibits equine infectious anemia virus Rev and feline immunodeficiency virus rev function but not the function of the hepatitis B virus posttranscriptional regulatory element. J Virol 72:7593–7597

Palacios I, Hetzer M, Adam SA, Mattaj IW (1997) Nuclear import of U snRNPs requires importin beta. EMBO J 16:6783–6792

Palmeri D, Malim MH (1996) The human T-cell leukemia virus type 1 posttranscriptional trans-activator Rex contains a nuclear export signal. J Virol 70:6442–6445

Paraskeva E, Izaurralde E, Bischoff FR, Huber J, Kutay U, Hartmann E, Lührmann R, Görlich D (1999) CRM1-mediated recycling of snurportin 1 to the cytoplasm. J Cell Biol 145:255–264

Pasquinelli AE, Powers MA, Lund E, Forbes D, Dahlberg JE (1997) Inhibition of mRNA export in vertebrate cells by nuclear export signal conjugates. Proc Natl Acad Sci USA 94:14394–14399

Plafker K, Macara IG (2000) Facilitated nucleocytoplasmic shuttling of the Ran binding protein RanBP1. Mol Cell Biol 20:3510–3521

Ribbeck K, Kutay U, Paraskeva E, Görlich D (1999) The translocation of transportin-cargo complexes through nuclear pores is independent of both Ran and energy. Curr Biol 9:47–50

Richards SA, Lounsbury KM, Carey KL, Macara IG (1996) A nuclear export signal is essential for the cytosolic localization of the Ran binding protein, RanBP1. J Cell Biol 134:1157–1168

Richards SA, Carey KL, Macara IG (1997) Requirement of guanosine triphosphate-bound ran for signal-mediated nuclear protein export. Science 276:1842–1844

Rittinger K, Budman J, Xu J, Volinia S, Cantley LC, Smerdon SJ, Gamblin SJ, Yaffe MB (1999) Structural analysis of 14-3-3 phosphopeptide complexes identifies a dual role for the nuclear export signal of 14-3-3 in ligand binding. Mol Cell 4:153–166

Rosorius O, Reichart B, Kratzer F, Heger P, Dabauvalle MC, Hauber J (1999) Nuclear pore localization and nucleocytoplasmic transport of eIF-5A: evidence for direct interaction with the export receptor CRM1. J Cell Sci 112:2369–2380

Roth J, Dobbelstein M, Freedman DA, Shenk T, Levine AJ (1998) Nucleo-cytoplasmic shuttling of the hdm2 oncoprotein regulates the levels of the p53 protein via a pathway used by the human immunodeficiency virus rev protein. EMBO J 17:554–564

Ruhl M, Himmelspach M, Bahr GM, Hammerschmid F, Jaksche H, Wolff B, Aschauer H, Farrington GK, Probst H, Bevec D, et al. (1993) Eukaryotic initiation factor 5 A is a cellular target of the human immunodeficiency virus type 1 Rev activation domain mediating trans-activation. J Cell Biol 123:1309–1320

Saitoh H, Sparrow DB, Shiomi T, Pu RT, Nishimoto T, Mohun TJ, Dasso M (1998) Ubc9p and the conjugation of SUMO-1 to RanGAP1 and RanBP2. Curr Biol 8:121–124

Sandri-Goldin RM (1998) ICP27 mediates HSV RNA export by shuttling through a leucine-rich nuclear export signal and binding viral intronless RNAs through an RGG motif. Genes Dev 12:868–879

Shah S, Tugendreich S, Forbes D (1998) Major binding sites for the nuclear import receptor are the internal nucleoporin Nup153 and the adjacent nuclear filament protein Tpr. J Cell Biol 141:31–49

Smitherman M, Lee K, Swanger J, Kapur R, Clurman BE (2000) Characterization and targeted disruption of murine Nup50, a p27-Kip1 interacting component of the nuclear pore complex. Mol Cell Biol 20:5631–5642

Solsbacher J, Maurer P, Bischoff FR, Schlenstedt G (1998) Cse1p is involved in export of yeast importin alpha from the nucleus. Mol Cell Biol 18:6805–6815

Stade K, Ford CS, Guthrie C, Weis K (1997) Exportin 1 (Crm1p) is an essential nuclear export factor. Cell 90:1041–1050

Stommel JM, Marchenko ND, Jimenez GS, Moll UM, Hope TJ, Wahl GM (1999) A leucine-rich nuclear export signal in the p53 tetramerization domain: regulation of subcellular localization and p53 activity by NES masking. EMBO J 18:1660–1672

Stutz F, Izaurralde E, Mattaj IW, Rosbash M (1996) A role for nucleoporin FG repeat domains in export of human immunodeficiency virus type 1 Rev protein and RNA from the nucleus. Mol Cell Biol 16:7144–7150

Taagepera S, McDonald D, Loeb JE, Whitaker LL, McElroy AK, Wang JY, Hope TJ (1998) Nuclear-cytoplasmic shuttling of C-ABL tyrosine kinase. Proc Natl Acad Sci USA 95:7457–7462

Taura T, Schlenstedt G, Silver PA (1997) Yrb2p is a nuclear protein that interacts with Prp20p, a yeast Rcc1 homologue. J Biol Chem 272:31877–31884

Taura T, Krebber H, Silver PA (1998) A member of the Ran-binding protein family, Yrb2p, is involved in nuclear protein export. Proc Natl Acad Sci USA 95:7427–7432

Toone WM, Kuge S, Samuels M, Morgan BA, Toda T, Jones N (1998) Regulation of the fission yeast transcription factor Pap1 by oxidative stress: requirement for the nuclear export factor Crm1 (Exportin) and the stress-activated MAP kinase Sty1/Spc1. Genes Dev 12:1453–1463

Toyoshima F, Moriguchi T, Wada A, Fukuda M, Nishida E (1998) Nuclear export of cyclin B1 and its possible role in the DNA damage- induced G2 checkpoint. EMBO J 17:2728–2735

Traglia HM, O'Connor JP, Tung KS, Dallabrida S, Shen WC, Hopper AK (1996) Nucleus-associated pools of Rna1p, the *Saccharomyces cerevisiae* Ran/TC4 GTPse activating protein involved in nucleus/cytosol transit. Proc Natl Acad Sci USA 93:7667–7672

Vetter IR, Arndt A, Kutay U, Görlich D, Wittinghofer A (1999) Structural view of the Ran-Importin beta interaction at 2.3 A resolution. Cell 97:635–646

Wada A, Fukuda M, Mishima M, Nishida E (1998) Nuclear export of actin: a novel mechanism regulating the subcellular localization of a major cytoskeletal protein. EMBO J 17:1635–1641

Watanabe M, Fukuda M, Yoshida M, Yanagida M, Nishida E (1999) Involvement of CRM1, a nuclear export receptor, in mRNA export in mammalian cells and fission yeast. Genes Cells 4:291–297

Wen W, Meinkoth JL, Tsien RY, Taylor SS (1995) Identification of a signal for rapid export of proteins from the nucleus. Cell 82:463–473

Wolff B, Sanglier JJ, Wang Y (1997) Leptomycin B is an inhibitor of nuclear export: inhibition of nucleo-cytoplasmic translocation of the human immunodeficiency virus type 1 (HIV-1) Rev protein and Rev-dependent mRNA. Chem Biol 4:139–147

Wu J, Matunis MJ, Kraemer D, Blobel G, Coutavas E (1995) Nup358, a cytoplasmically exposed nucleoporin with peptide repeats, Ran- GTP binding sites, zinc fingers, a cyclophilin A homologous domain, and a leucine-rich region. J Biol Chem 270:14209–14213

Yan C, Lee LH, Davis LI (1998) Crm1p mediates regulated nuclear export of a yeast AP-1-like transcription factor. EMBO J 17:7416–7429

Yang J, Bardes ES, Moore JD, Brennan J, Powers MA, Kornbluth S (1998) Control of cyclin B1 localization through regulated binding of the nuclear export factor CRM1. Genes Dev 12:2131–2143

Yokoyama N, Hayashi N, Seki T, Panté N, Ohba T, Nishii K, Kuma K, Hayashida T, Miyata T, Aebi U, et al. (1995) A giant nucleopore protein that binds Ran/TC4. Nature 376:184–188

Zhang MJ, Dayton AI (1998) Tolerance of diverse amino acid substitutions at conserved positions in the nuclear export signal (NES) of HIV-1 Rev. Biochem Biophys Res Commun 243:113–116

Zolotukhin AS, Felber BK (1997) Mutations in the nuclear export signal of human ran-binding protein RanBP1 block the Rev-mediated posttranscriptional regulation of human immunodeficiency virus type 1. J Biol Chem 272:11356–11360

Zolotukhin AS, Felber BK (1999) Nucleoporins nup98 and nup214 participate in nuclear export of human immunodeficiency virus type 1 Rev. J Virol 73:120–127

Structures of Importins

Elena Conti[1]

1 Introduction

Importins are nuclear transport receptors of the karyopherin superfamily. They recognize nuclear proteins that are synthesized in the cytoplasm and translocate them to the nucleus (Mattaj and Englmeier 1998; Weis 1998; Görlich and Kutay 1999; Nakielny and Dreyfuss 1999). Macromolecules are targeted to the nucleus by a signal sequence that mediates the interaction with an importin. The founding member of the karyopherin superfamily is the importin-αβ heterodimer, which binds and transports proteins containing classical, positively-charged nuclear localization signals. The α component of the heterodimer (known in the literature as importin-α, karyopherin-α, hSRP1 or p58) functions as an adapter by binding the nuclear localization signal (NLS)-containing protein and the carrier component β (Fig. 1). The β-component (importin-β, karyopherin-β1, p97) docks the complex to distinct sites along the fibrils of the nuclear pore complex (NPC). At the nucleoplasmic side, β releases its cargo by interacting with the small GTPase Ran in its GTP-bound form. The NLS-containing protein diffuses within the nucleus to exert its function, while the soluble components of the transport machinery are recycled back to the cytoplasm for a new import cycle. The β-RanGTP complex can exit the nucleus directly, while the adapter α requires a specific exportin of the karyopherin-β superfamily to be translocated out the nucleus. At the cytoplasmic side, the GTPase loses its high affinity for β as it is hydrolyzed to the GDP-bound form by the combined action of a RanBD (Ran binding domain) and RanGAP (see Chap. 3 by Bischoff and references therein). RanGDP is transported back into the nucleus via an association with the transport factor NTF2, and there converted to the GTP-bound form by the action of a specific GEF (the nuclear guanine nucleotide exchange factor RCC1). The asymmetric distribution of cytoplasmic RanGDP and nuclear RanGTP drives the directionality of transport processes, ensuring cargo loading and cargo release in the appropriate cell compartments.

Many nuclear proteins are transported via the NLS-αβ import pathway. Despite being considered the classical import pathway for historical reasons,

[1] EMBL, Structure Program, Meyerhofstrasse 1, 69117 Heidelberg, Germany

Results and Problems in Cell Differentiation, Vol. 35
K. Weis (Ed.): Nuclear Transport

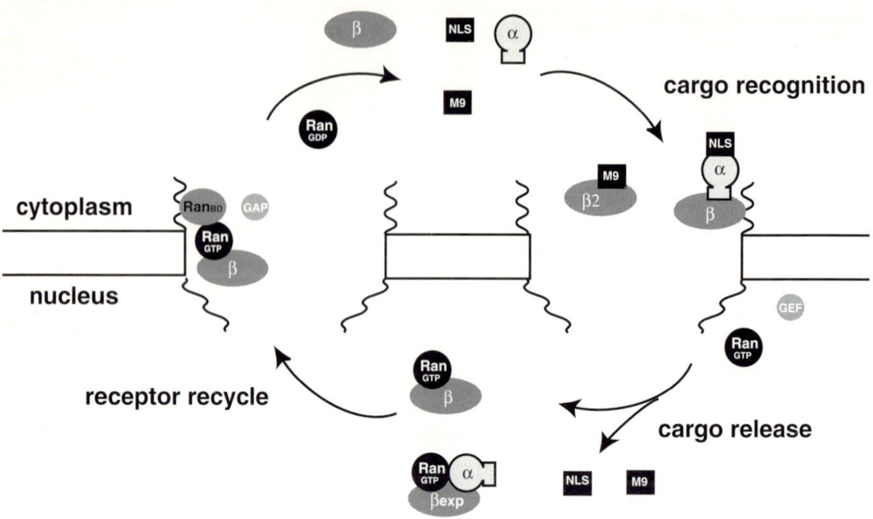

Fig. 1. Simplified scheme of nuclear import. In the cytoplasm, the nuclear targeting signal of a macromolecular cargo binds to its cognate nuclear transport receptor, either directly (for example the M9 signal to β2) or indirectly (for example the NLS signal to β via the adapter molecule α). The β receptor docks to the fibrils of the nuclear pore and translocates the cargo to the nucleus. At the nucleoplasmic side, RanGTP binds to β releasing the cargo. The β receptor is recycled to the cytoplasm, where the action of RanBD and GAP efficiently converts Ran to its GDP form. α is recycled to the cytoplasm via a specific exportin

the requirement of an adapter molecule is in fact not universal. Different import receptors of the karyopherin-β superfamily are known to transport ribosomal proteins and mRNA binding proteins by fulfilling both the signal-binding and NPC-docking activities. For example, β2 (karyopherin-β2 or transportin) binds the M9 signal of its cargo directly (Fig. 1). Analogously, β can transport certain proteins by direct recognition of a signal that is similar to the importin-β-binding (IBB) domain of α and that is also present in another adapter molecule, snurportin. Strikingly, the IBB domain of α that binds to β is somewhat similar to an NLS, with conserved positive charges interspersed along the sequence. Why cannot classical NLSs bind directly to β? What defines classical NLSs, given their variability but also the discrimination of the import machinery against apparently similar sequences that do not have import activity? How does RanGTP trigger the unloading of the cargo in the nucleus and why is RanGDP unable to do so in the cytosol? How does β dock to the nuclear pore and why does RanGTP disrupt this interaction to end an import translocation event? Understanding the molecular mechanisms that govern this complex series of events requires structural knowledge of the proteins involved and their changes as they interact with different partners in the import cycle.

Crystal structures of several nuclear import components have been determined to date (Fig. 2). X-ray crystallography is a powerful technique to obtain structural information at nearly the atomic level, but it does have limitations. While a macromolecule in solution might undergo significant motion, it needs to be locked in a single conformational state to be studied by X-ray crystallography. In the case of flexible multidomain and multifunctional proteins, often the molecule has to be dissected into fragments to be efficiently locked in a single state and crystallized. A prerequisite of any crystal structure determination is the arrangement of a large number of identical molecules in a periodic crystal lattice, and the outcome is the determination of their average atomic positions. As a result, the average position of the atoms is more accurately defined in rigid parts of the molecule than in those with more pronounced vibrational motions. The extreme cases are totally flexible parts of the molecule (such as surface loops that are not restrained by specific interactions) which appear disordered as they assume many different conformations in the solvent channels of the crystal. Crystal structures may also be less accurate at lattice contacts where the conformation of surface residues can be distorted, and are less precise at low resolution. A low-resolution 3.0Å crystal structure does not reveal the role of water molecules nor the precise hydrogen-bonding pattern in an enzymatic catalytic mechanism, but it is sufficient to show the details of the macromolecular interactions underpinning a protein-protein recognition event.

2 NLS Recognition: Association with α

Nuclear localization signals lack a strict consensus sequence but are, in general, short stretches characterized by a high proportion of positively-charged residues (Dingwall and Laskey 1991). The basic amino acids are grouped either in a single cluster, as in the case of the monopartite SV40 T antigen NLS (Kalderon et al. 1984; Lanford and Butel 1984), or in two clusters, as in the case of the bipartite nucleoplasmin NLS (Dingwall et al. 1982). The mere juxtaposition of basic residues is, however, not sufficient to determine whether a sequence functions as a nuclear targeting signal, the specific position and character of the residues in the sequence being essential. A single-point mutation within the five contiguous positively-charged residues of the SV40 T antigen NLS has no effect when the modified residue is the arginine at P4 (Fig. 3A), while nuclear targeting is abrogated when the lysine at P2 is mutated (Colledge et al. 1986). The basic cluster of the SV40 T antigen NLS resembles the downstream cluster of bipartite NLSs, which in nucleoplasmin consists of four contiguous lysines (Fig. 3A). Despite the similarity, the downstream cluster of nucleoplasmin is not sufficient to target a protein to the nucleus and an upstream cluster of two positively charged residues is required. The linking peptide can have a variable sequence but must be at least ten residues in length (Dingwall et al. 1988; Robbins et al. 1991). Puzzling in this context is the NLS

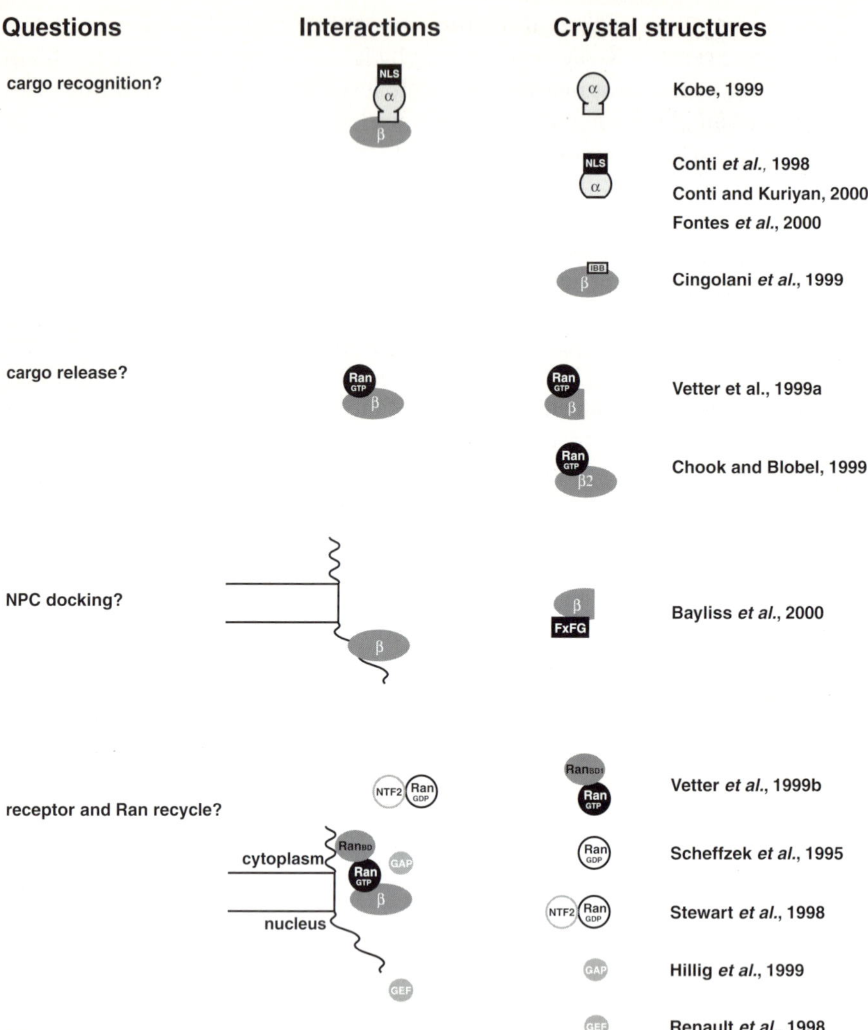

Fig. 2. Aspects that have been addressed to date by X-ray crystallography include the structures of: full-length importin-/karyopherin–α (Kobe et al. 1999); the nuclear localization sequence (NLS) binding domain of α bound to several NLSs (Conti et al. 1998; Conti and Kuriyan 2000; Fontes et al. 2000); importin-β bound to the IBB domain of α (Cingolani et al. 1999);fragment of β bound to RanGTP (Vetter et al., 1999a); tranportin-/karyopherin-β 2 bound to RanGTP (Chook and Blobel 1999); fragment of β bound to nucleoporin Phe-x-Phe-Gly (FxFG) repeats (Bayliss et al. 2000); a Ran binding domain (RanBD1) of RanPB2 bound to RanGTP (Vetter et al. 1999b); RanGDP (Scheffzek et al. 1995); RanGDP bound to NTF2 (also known as p10) (Stewart et al. 1998); Ran GTPase activating protein (RanGAP; Hillig et al. 1999); Ran guanidine exchange factor (RanGEF, also known as RCC1; Renault et al. 1998)

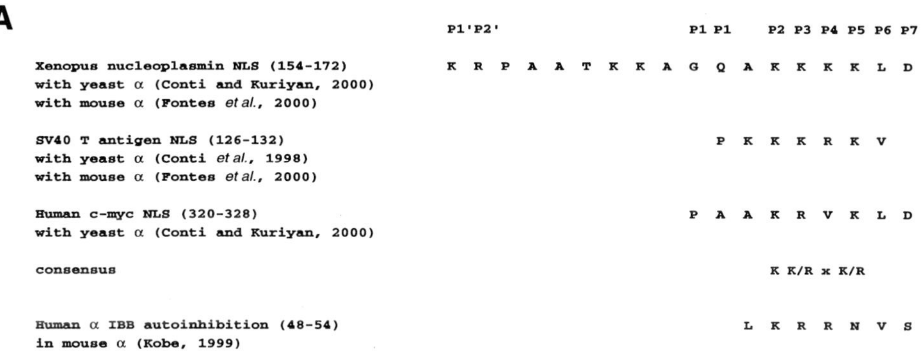

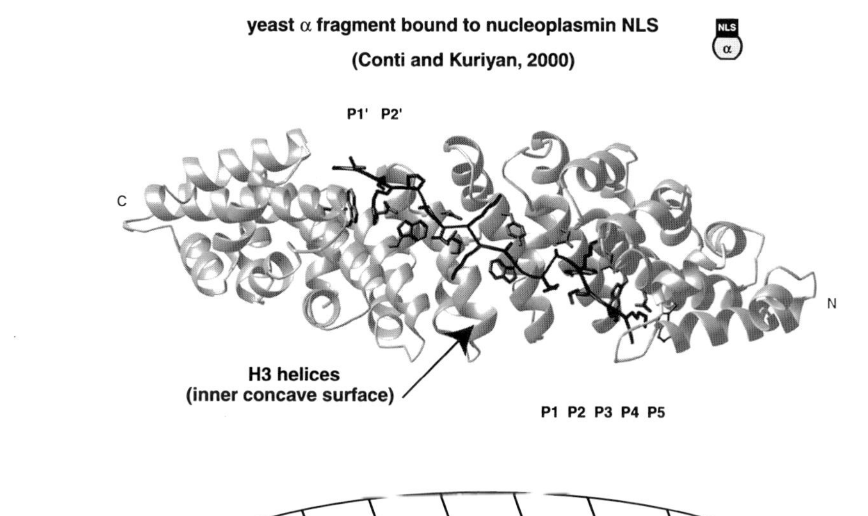

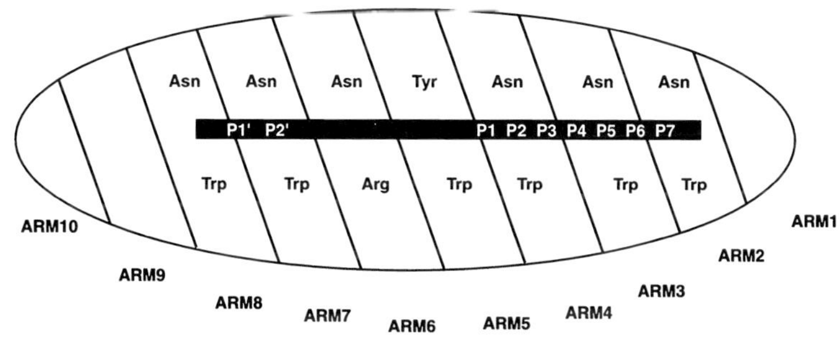

Fig. 3A–C. NLS recognition by karyopherin-/importin–α. **A** Structure-based sequence alignment of bipartite and monopartite NLSs. The corresponding P and P′ pockets for NLS residues are indicated. Part of the IBB domain binds similarly to a NLS at the P pockets. **B** The structure of the NLS-binding fragment of yeast α in *gray* (88–530) bound to a peptide containing the nucleoplasmin NLS sequence (in *black*). The peptide binds in an extended conformation with the upstream cluster of positive charges at the P′ binding pockets on the protein and the larger downstream cluster at the P sites. The side chains of conserved residues of α that interact with the two clusters and with the linker of this bipartite NLS are shown in *darker gray*. **C** The architecture of the NLS-binding domain of α, which is built by the tandem packing of 10 ARM repeats. The central repeats line the surface of the molecule with mostly Trp and Asn residues which shape the P and P′ binding pockets

of c-*myc*, which is characterized by even fewer positively charged residues than a typical downstream cluster of bipartite NLSs (Fig. 3A) but which nevertheless functions as an effective monopartite signal (Dang and Lee 1988; Makkerh et al. 1996). The very basic SV40 T antigen NLS, the more hydrophobic c-*myc* NLS and the longer nucleoplasmin NLS are all specifically recognized by the adapter importin-/karyopherin-α.

The adapter α is a 60 kDa molecule composed of a 50 kDa NLS-binding domain connected by a flexible linker to a short N-terminal IBB domain (Görlich et al. 1996a; Weis et al. 1996). The structure of the NLS-binding domain of the yeast (Conti et al. 1998) and mouse (Kobe 1999) proteins are essentially identical, reflecting the high degree of conservation between lower and higher eukaryotes (44% sequence identity between yeast and mouse α). The NLS-binding domain is an entirely α-helical molecule (Fig. 3B) built from the tandem stacking of ten repeating units known as ARM motifs. These consist of about 40 amino acid residues (Peifer et al. 1994) which fold into three α-helices (H1, H2 and H3) that are arranged with an approximately triangular cross-section. The repeats pack side-by-side in an almost parallel fashion with a 30° rotation between contiguous units, creating an elongated molecule with an overall superhelical twist. The roughly parallel H3 helices of the repeats form the concave surface of the molecule, which presents an elongated and shallow surface groove. The groove is lined with conserved residues, in particular with an array of mostly tryptophan residues each at an identical position on a given H3 helix, and with a parallel array of mostly asparagine residues one turn of a helix downstream (Fig. 3B). The repeated architecture of α results in a regular ladder of conserved Trp-Asn pairs, which is interrupted in the central part of the molecule (ARM 5 and ARM 6) by two different but still invariant residues (Fig. 3 C).

The structures of yeast and mouse α in complex with the NLS sequences of nucleoplasmin, SV40 T antigen and c-*myc* show a consensus mode of chemical recognition (Conti et al. 1998; Conti and Kuriyan 2000; Fontes et al. 2000). The NLSs bind in an extended conformation along the floor of the conserved surface groove, as shown in the case of the bipartite signal in Fig. 3B. The groove is shaped with an array of binding pockets that are formed by the multiple Trp-Asn pairs and function as recognition motifs for the NLS Lys and Arg residues (Fig. 4). The Asn residues hydrogen-bond to the NLS backbone anchoring it in an extended conformation. The ladder of tryptophan side chains creates regularly-spaced hydrophobic pockets that enclose the aliphatic portion of lysine residues of the NLS on one side of the groove. The positively-charged tip of the lysine side chains of the NLS are recognized by electrostatic interactions with conserved negatively-charged and polar residues at the immediate periphery. The number of Lys or Arg residues potentially recognized by α is related to the number of recognition motifs present on the surface. A larger binding site (P) is present in the N-terminal half of the molecule (3 Trp-Asn pairs), while a smaller binding site (P′) is present in the C-terminal half (2 Trp-Asn pairs).

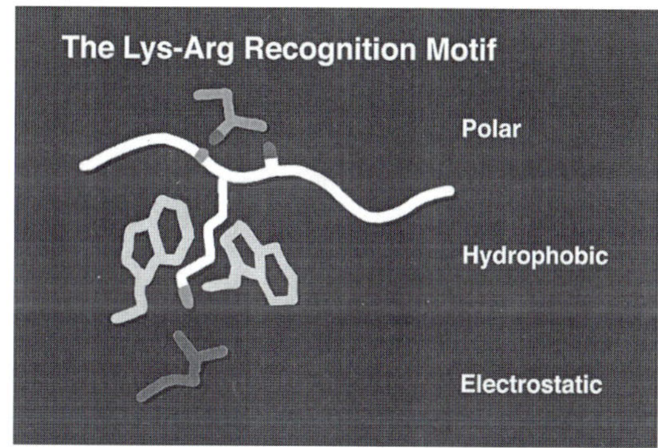

Fig. 4. The interactions responsible for the specific recognition of several NLS lysine (or arginine) residues. Asn residues of the protein are involved in polar interactions with the backbone of the NLS, Trp residues in hydrophobic interactions with the aliphatic portion of the Lys side chain, and acidic residues in electrostatic interactions with the positively charged tip of the Lys side chain

The nucleoplasmin bipartite NLS binds with its 2-Lys upstream cluster at the smaller binding site and its 4-Lys downstream cluster at the larger binding site (Fig. 3). Positively charged residues of the clusters intercalate between the Trp-Asn pairs. The two clusters are connected by a linker, which nestles along the surface groove at the central ARM repeats. Here the conserved residues interrupting the regularity of the Trp-Asn pairs are found in yeast α to anchor the linker by hydrogen-bonding to its main chain. The side chains of the linker are not involved in specific interactions with the protein, explaining the variability of the linker sequence in bipartite NLSs. The length of the linker is important in positioning the two clusters correctly, as it would not be possible for a shorter intervening sequence to span the physical separation between the P′ and P sites.

The SV40 T antigen and c-*myc* monopartite NLSs bind at the larger binding site (P) where the downstream cluster of the bipartite nucleoplasmin NLS is recognized (Fig. 3A). Consistent with mutagenesis data, the lysine residue at P2 is involved in the tightest electrostatic interactions, while no charged or polar residues surround the P4 pocket, where an Arg in the SV40 T antigen is substituted by a Val in c-*myc*. The combination of hydrophobic and electrostatic properties at the P2, P3 and P5 pockets is consistent with the K-K/R-x-K/R (Lys-Lys/Arg-x-Lys/Arg) consensus proposed for monopartite NLSs (Chelsky et al. 1989). At P1, residues that are compatible with sharp turns (Pro and Gly) are best suited to the curvature of the pocket, while a small hydrophobic and an acidic residue are preferred at P6 and P7, respectively. Monopartite NLSs are engaged in more extensive interactions than the downstream cluster of the bipartite NLS, even though they bind at the same pockets. Monopartite NLSs therefore represent a very efficient variant of a bipartite NLS downstream cluster, where suboptimal interactions at the large binding site are tolerated due to the abundance of the contacts made along the whole surface groove by the upstream cluster and the linker backbone. The presence of multiple binding pockets in the adapter α explains the lack of a strict consensus in NLS sequences and results in the versatility of NLS recognition. The NLS is presumably molded into an extended conformation upon binding to α. The monopartite NLS of NFκB, for example, is flexible and unstructured in the nuclear form of the transcription factor when bound to DNA (Müller et al. 1995), and is in a helical conformation in the cytoplasmic form when bound to IkBα (Jacobs and Harrison 1998).

3 Cargo Binding to the Receptor: Importin-β-Binding Association with β

The adapter α is able to bind NLSs, but it binds them tighter when in complex with importin-/karyopherin-β (Moroianu et al. 1996). The N-terminal domain of α that interacts with β is known as the IBB domain and is a 44-residue-long stretch rich in arginines and lysines (Görlich et al. 1996a; Weis et al. 1996). In the absence of β or an NLS, a small part of the IBB functions as an autoinhibitory segment, binding at the larger NLS binding site of α (Kobe 1999). The binding affinity is very low in vitro (Görlich et al. 1996a; Weis et al. 1996) and in fact the sequence of the autoinhibitory part of the IBB is not optimal for binding at the NLS recognition site as it lacks a basic residue at P5 and has less suitable amino-acid side chains at P1 and P7 (Fig. 3A). Synergistic binding of the autoinhibitory segment is probably achieved by the covalent attachment of the IBB to the NLS binding domain, which effectively increases the local concentration of the two domains. Apart from the small autoinhibitory segment, most of the IBB domain is disordered and presumably flexible in the structure of full-length mouse α. Upon formation of the αβ heterodimer, the entire IBB domain of α, including its autoinhibitory segment, interacts with β.

The structure of β in complex with the IBB domain of α (Fig. 5A) shows that the β-receptor is a superhelical molecule with its C-terminal half tightly wrapped around the cargo (Cingolani et al. 1999). The β-receptor is built of 19 HEAT repeats, each consisting of two helices (A and B). The A and B helices are joined by a turn that is typically short, with the exception of a longer intra-repeat acidic loop at HEAT 8. The inter-repeat connections are linkers with variable lengths and variable conformations, giving rise to a relatively irregular shape as compared with the regular HEAT-repeat protein phosphatase 2A (Groves et al. 1999). The HEAT repeats have rather variable sequences but a few hydrophobic residues are present at conserved positions (Andrade and Bork 1995). The periodicity of conserved residues of the HEAT repeats of β resembles the consensus sequence of the ARM repeats of α (Fig. 5B). Indeed, the structures of individual HEAT and ARM motifs are very similar. When the structures of the two motifs are superposed, the B helix of the HEAT motif overlays the H3 helix of the ARM structure, while a part of the A helix of the

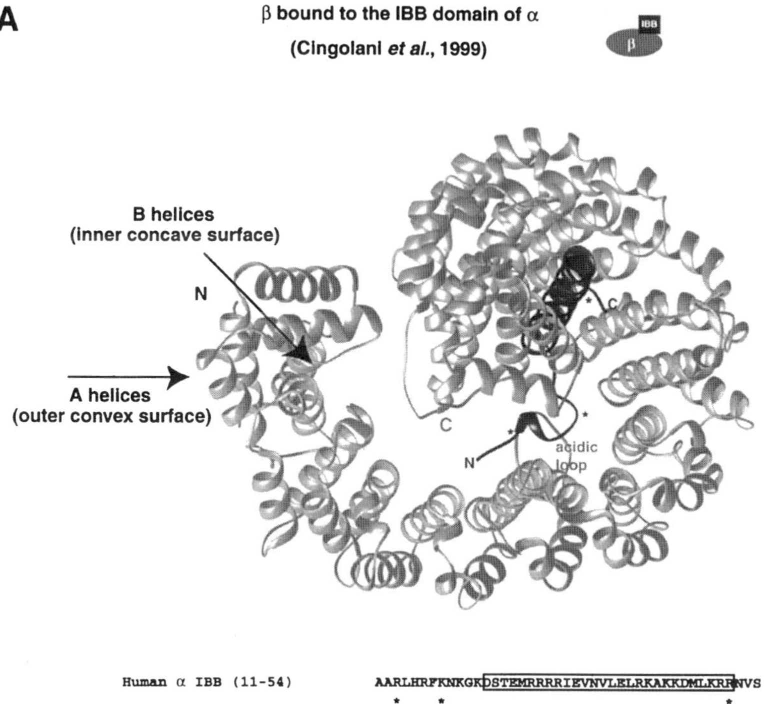

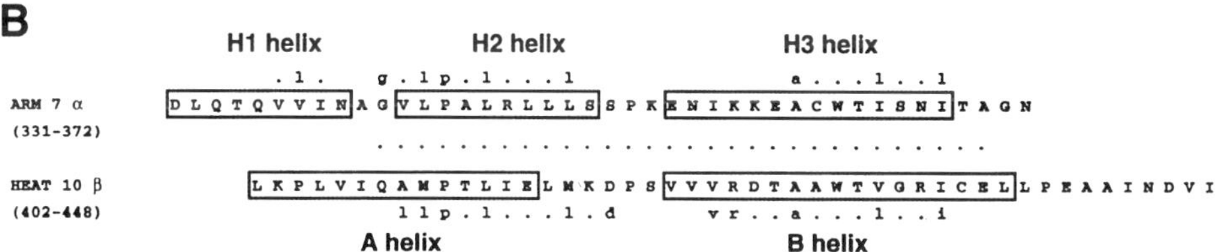

Fig. 5A,B. Cargo recognition by karyopherin-/importin-β. A Structure of full-length β (in *gray*) bound to the IBB domain of α (in *black*). The IBB cargo is wrapped by the inner B helices of the C-terminal half of β. The N- and C-termini of the two molecules are indicated, as well as the position of the conserved acidic loop of β. Most of the basic IBB domain folds into a helical conformation (*boxed* residues) when bound to β, with three key residues involved in specific interaction (indicated by the *asterisk*). B Structure-based sequence alignment of a prototypical ARM repeat of α and of a HEAT repeat of β. The *boxes* enclose helical residues and the consensus for ARM and HEAT repeats is shown in *lower case letters*. Superposition of the two repeats reveals striking structure similarity, with the Cα atoms of most residues being within 1.5 Å (as indicated with *dots*)

HEAT motif overlays the H2 helix (Fig. 5B). The main difference between the two types of repeat is a kink in helix A of the HEAT motifs that is induced by the presence of a proline residue. This kink is more pronounced in ARM motifs and results in the generation of two different helices (H1 and H2) instead of a bent one. Analogous to the architecture of α, the HEAT repeats of β stack side-by-side generating a superhelix with the inner concave surface formed by B helices. Despite the topological similarity, the overall shape of the α and β superhelices differ as a consequence of the different number of repeats and of the different packing interactions of their N-termini. Moreover, circumstantial evidence suggests that the presence of the IBB cargo is likely to affect the overall conformation. The increased susceptibility of β to proteases in the absence of the IBB cargo points to a less compact conformation in the free state than the observed snail-like conformation of the cargo-bound state (Cingolani et al. 1999).

Upon binding to β, the IBB domain folds into an N-terminal-extended portion and a long helix (Fig. 5A). The IBB helix is bound by HEAT repeats 12–19 which spiral tightly around the helix. The basic IBB helix is bound by electrostatic complementarity, with several of its positively-charged residues engaged in ionic interactions with aspartic and glutamic acid residues that line the eight C-terminal inner B helices of β. The small N-terminal moiety of the IBB domain is bound in an extended conformation between HEAT 7 and 11, making extensive interactions with the acidic loop. The mode of recognition of a few key positively-charged residues resembles the chemical recognition of basic NLSs by α. Not only the positively-charged tips of the Lys/Arg side chains are held in position through electrostatic interactions with negatively charged residues, but also the aliphatic portions of their side chains are involved in hydrophobic interactions with tryptophan residues. A tryptophan residue in the conserved acidic loop (Trp 342) and a tryptophan residue on the B helix of HEAT 10 (Trp 430) interact with basic residues of the IBB N-terminal portion that are invariant across species (Arg 13 and Lys 18). The same chemical recognition is present at the very C-terminus of the IBB domain, with Arg 51 packing against a tryptophan residue of helix 19B. The solvent-exposed tryptophan residues on the B helices of the HEAT motifs of β are located at identical positions as the tryptophan residues in the H3 helices of the ARM motifs of α involved in NLS binding (Fig. 5B), making the suggestion that α and β are evolutionarily related (Malik et al. 1997) all the more compelling. Despite being built by similar repeating units (ARM and HEAT), despite binding positively-charged residues with a similar strategy and despite employing tryptophan residues at identical positions in the repeats, α and β have different sequence requirements for their ligands.

The IBB cargo is bound to β by specific chemical recognition of a few key residues at its N- and C-termini, and with the intervening basic helix being involved in additional multiple ion-pairing interactions. As in the case of α-NLS recognition, the combination of both hydrophobic and electrostatic interactions on the surface of the molecule results in binding of positively charged amino acids with high affinity and specificity. The positions of these structural determinants on the surface of the receptor determine the spacing of essential Lys and Arg residues in the sequence of the IBB cargo and thus discriminate against other potential positively-charged substrates. Classical positively-charged NLSs that bind α are too short and have basic residues at positions that are not consistent with the spacing required by β. Even the mode of binding of the same C-terminal portion of the IBB domain (residues 48–54) to α and to β is different, and is molded on the characteristics of the molecules. When bound to β, the sequence adopts a helical conformation with Arg 51 making essential interactions (Cingolani et al. 1999), while when bound and autoinhibiting α it stretches in an extended conformation with Arg 51 occupying the non-essential P4 position (Kobe 1999). The interaction of β with the IBB domain of the NLS-binding adapter α is analogous to a direct β-cargo interaction when no adapter is needed, for example in the case of the direct recognition of the nuclear targeting signals of HIV Rev and Tat (Henderson and Percipalle 1997; Truant and Cullen 1999). Although sharing sequence similarity, the NLSs of HIV Rev and Tat are, however, shorter than the IBB domain, and their precise mode of recognition is unknown at present.

4 Cargo Release: RanGTP Binding to β and β2

Once in the nucleus, the β-cargo complex is dissociated by RanGTP. Like all other Ras-related GTPases, Ran has a core structure known as the G domain featuring two loop or switch regions that can adopt different conformations. When GTP is bound, the switch I and switch II regions move closer to the nucleotide to interact with its γ-phosphate and assume a conformation that is typical of all Ras-related proteins in the "on" state. When GDP is bound, the switch regions move away from the nucleotide and assume a conformation that is Ran-specific. In addition to the common G domain, Ran has a C-terminal tail of about 40 residues that functions as an extra switch responding to the nucleotide state of the GTPase (Fig. 6). In the GDP-bound form, the C-terminal tail forms an extended portion and a contiguous α-helix that pack on the G-domain (Scheffzek et al. 1995). The very C-terminal acidic DEDDDL sequence is not ordered in the crystals but is likely to be involved in general electrostatic interactions with a basic patch present on the Ran surface (residues 139–142), 35Å from the β-phosphate of GDP (Fig. 6D). The C-terminal tail is flipped away from the G-domain in the structures of the GTP-bound form of Ran in complex with three effectors, the RanGTP-binding proteins β, β2 and RanBD (Fig. 6A–C). The extended portion of the tail in the conformation observed in the GDP-bound form is sterically incompatible with the switch I region conformation adopted in the GTP-bound form. A similar steric clash is predicted to occur in uncomplexed RanGTP, for which no structure is available at present, although in this case biochemical studies suggest that the acidic moiety might still be attached to the G domain (Macara 1999).

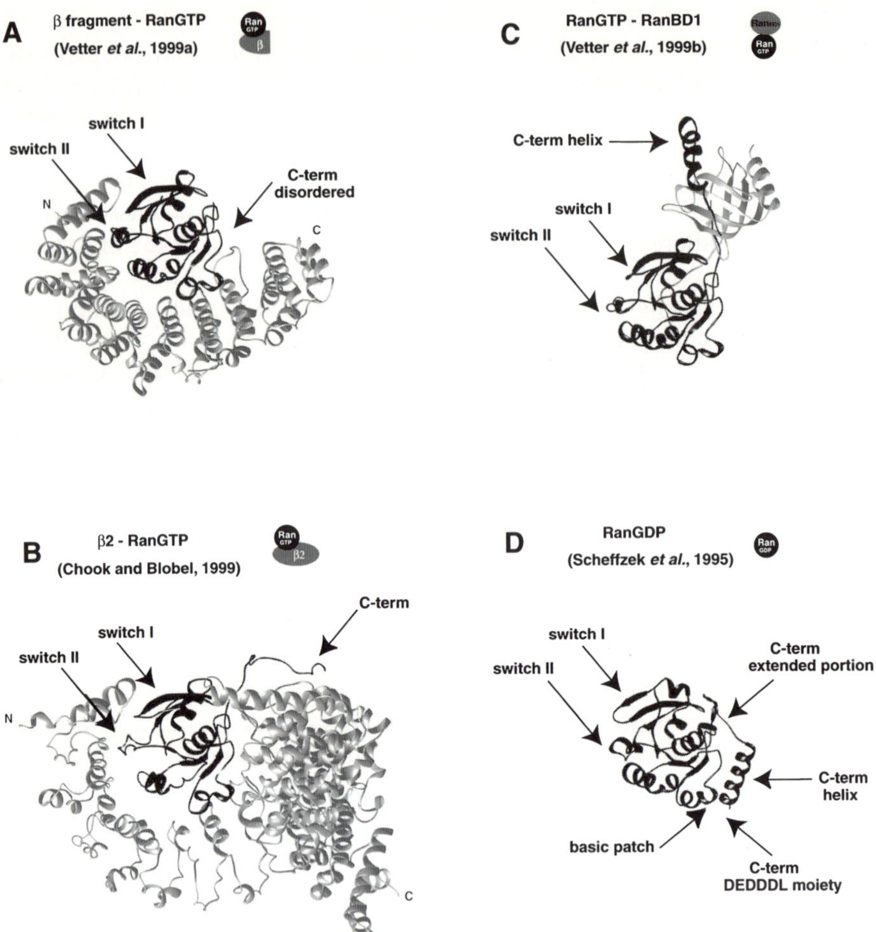

Fig. 6A–D. Structures of Ran. RanGTP is known in complexes with three effectors: **A** with a fragment of β (HEAT 1–10) shown in a similar orientation as in Fig. 5A; **B** with full-length β2; and **C** with RanBD1. **D** Structure of cytoplasmic RanGDP. The structures are shown with Ran in a similar orientation. The positions of the switch regions of Ran (switch I, switch II, and C-terminal tail) are indicated

The structures of RanGTP bound to a fragment of β comprising the first ten HEAT repeats (Vetter et al. 1999a) and of that bound to full-length β2 (Chook and Blobel 1999) show that the GTPase binds at the inner concave surface of the N-terminal arch of the importin molecules (Fig. 6). A comparison of the structure of the first 10 HEAT repeats of β and β2 reveals that the N-terminal arch of the β receptors has a similar crescent shape, independently of whether they are bound to cargo or to RanGTP (Figs. 5A, 6A,B). The C-terminal arch of the receptors, however, has a different orientation in the two full-length structures known to date. While β bound to its cargo has a closed conformation with the C-terminal arch oriented towards the N-terminus of the molecule (Fig. 5A), β2 bound to RanGTP has a more open conformation with the C-terminal arch almost orthogonal to the N-terminal one (Fig. 6B). Despite the substantial differences in overall tertiary structure, the receptors obey a very similar topological scheme, as indicated in the structural alignment shown in Fig. 7, obtained by superposing the N-terminal arch and the C-terminal arch of the two full-length structures independently. In the original papers (Fig. 7), different numbering of the repeats in the β and β2 structures is used. For clarity, the HEAT repeat numbering defined for β1 is used for both structures in this discussion.

The two β molecules use a similar shape-recognition mechanism to bind the nuclear form of the GTPase. The first HEAT repeat interacts with a small region of switch I, most of which, however, remains exposed to solvent (Figs. 6,7). The second HEAT repeat binds the switch II region, burying it with mainly hydrophobic interactions. While the interactions in the switch I region are moderate in comparison to switch II, the importance of its recognition is shown by the inhibitory effect on nuclear import upon the removal of the first 44 residues of β (Kutay et al. 1997). The basic patch on Ran binds to negatively-charged residues that line the 7B helix in the central part of the importin-β structures. At HEAT 8, the acidic loop interacts with the region of the G domain that is covered by the C-terminal helix of Ran in the GDP-bound form, but that is accessible for interaction in the GTP-bound form. The C-terminal portion of the acidic loop of both importins would spatially overlap with the C-terminal helix in RanGDP. The conformation of the C-terminal tail and of the switch regions in the GDP-bound of Ran would be incompatible with Ran's interaction in the crescent of β and β2. Despite these gross similarities that are responsible for the discrimination of β and β2 against RanGDP, the chemical details of the interactions with RanGTP are different in the two receptors and are tailored to their different amino-acid sequences. Perhaps the most striking topological difference between the two receptors is the acidic loop (Fig. 7). The acidic loop in β2 is characterized by a large insertion, the central part of which (residues 333–340) contacts Ran in a region that is not involved in contacts in the complex with the β fragment. Only a few of the interactions with Ran involve identical residues at identical positions in the two receptors, namely a leucine in helix 2B, a glutamic acid in 7B and a tryptophan residue in the acidic loop (Fig. 7).

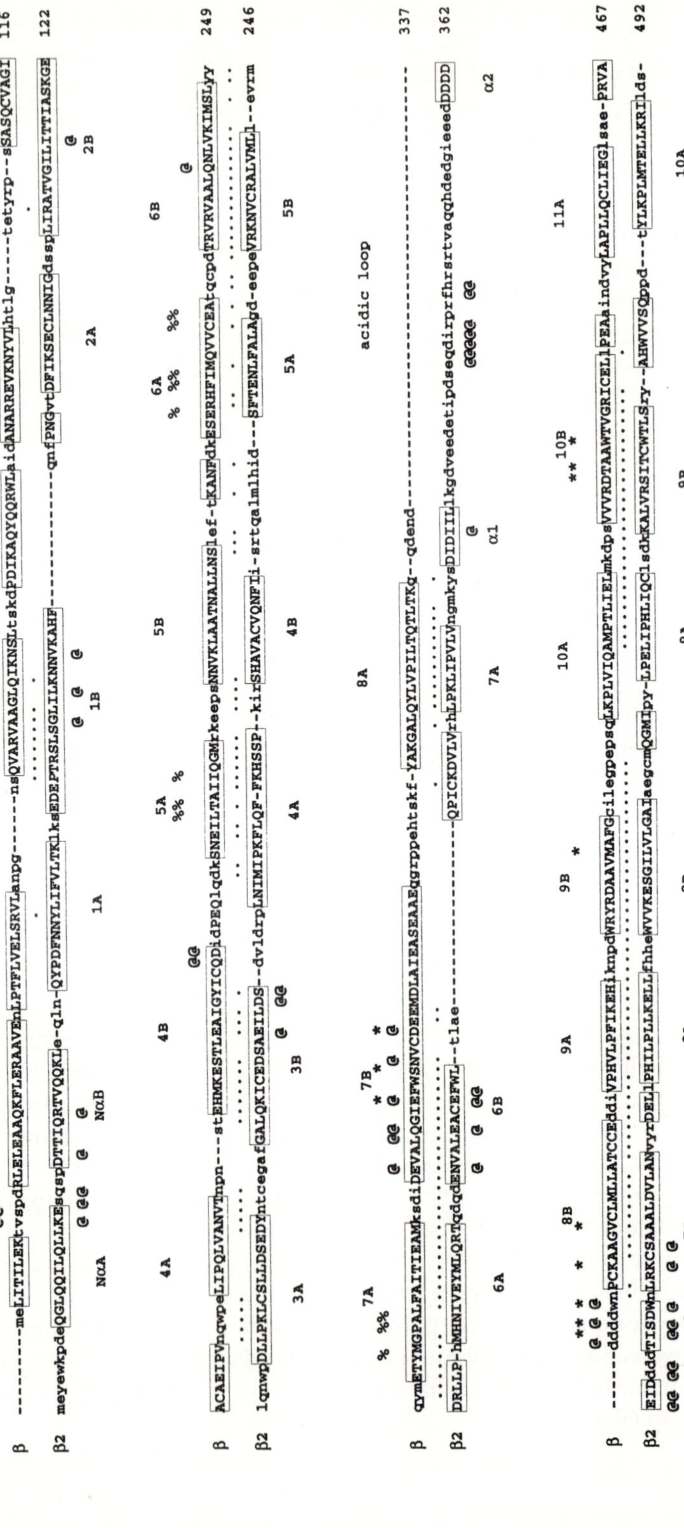

```
            1A              1B                2A                       2B                                3A                        3B
                @@                                                     @   @    @@                                                      @        @
β   ---------meLITILEKtvspdRLELEAAQKFLERAAVEnLPTFLVELSRVLanpg------nsQVARVAAGLQIKNSLtskdPDIKAQYQQRWLaidANARREVKNYVLhtlg-----tetyrp--sSASQCVAGI  116
                                                            .            ........  .                                                  .
β2  meyewkpdeQGLQQILQLLKEsqspDTTIQRTVQQKLe-qln-QYPDFNNYLIFVLTKlksEDEPTRSLSGLILKNNVKAHF-------------qnfPNGvtDFIKSECLNNIGdsspLIRATVGILITTIASKGE  122
                @ @@   @    @                                      @  @   @                                                @
            NαA             NαB             1A                      1B                             2A                    2B

            4A                 4B                  5A           5B                  6A            6B
                                                   %%  %                       %  %%    %%             @
                                  @@
β   ACAEIPVnqwpeLIPQLVANVTnpn---stEHMKESTLEAIGYICQDidPEQlqdkSNEILTAIIQGMrkeepsNNVKLAATNALLNSlef-tKANFdkESERHFIMQVVCEAtqcpdTRVRVAALQNLVKIMSLyy  249
         .....    .....    .......  ...  .         ..  .. .  ......  ....       ...   .    .  .    ..  .   .  .  ..  .............   . ..
β2  lqnwpDLLPKLCSLLDSEDYntcegafGALQKICEDSAEILDS--dvldrpLNIMIPKFLQF-FKHSSP--kirSHAVACVQNFIi-srtqalmlhid---SFTENLFALAgd-eepeVRKNVCRALVMLl--evrm  246
                                   @  @@
          3A                     3B                   4A              4B                   5A           5B

             7A            7B                               8A                                    acidic loop
      %  %%           *  *  *
               @  @@  @  @  @
β   qymETYMGPALFAITIEAMksdiDEVALQGIEFWSNVCDEEMDLAIEASEAAEqgrppehtskf-YAKGALQYLVPILTQTLTKq--qdend--------------------------------------------  337
     .......  ..........................  ..                    .    .  ...........   .
β2  DRLLP-hMHNIVEYMLQRTqdqdENVALEACEFWL--tlae------------------QPICKDVLVrhLPKLIPVLVngmkysDIDIILlkgdveedetipdseqdirprfhrsrtvaqqhdedgieeedDDDD  362
                          @   @  @@                                                        @                 @@@@@  @@
          6A               6B                                   7A               α1                                                          α2

            ** *
            @ @ @   *   *
                       8B           9A          9B  *               10A                 ** *10B                  11A
β   ------ddddwnPCKAAGVCLMLLATCCEddiVPHVLPFIKEHiknpdWRYRDAAVMAFGcilegpepsqLKPLVIQAMPTLIELmkdpsVVVRDTAAWTVGRICELlPEAaindvyLAPLLQCLIEGlsae-PRVA  467
                 .. ........ ........ .....................................             ..........................     .
β2  EIDdddTISDWnLRKCSAAALDVLANvyrDELlPHILPLLKELLfhheWVVKESGILVLGAIaegcmQGMIpy-LPELIPHLIQClsdkKALVRSITCWTLSry--AHWVVSQppd---tYLKPLMTELLKRIlds-  492
    @@ @@  @@ @  @ @
            α3       7B          8A           8B                   9A             9B                       10A
```

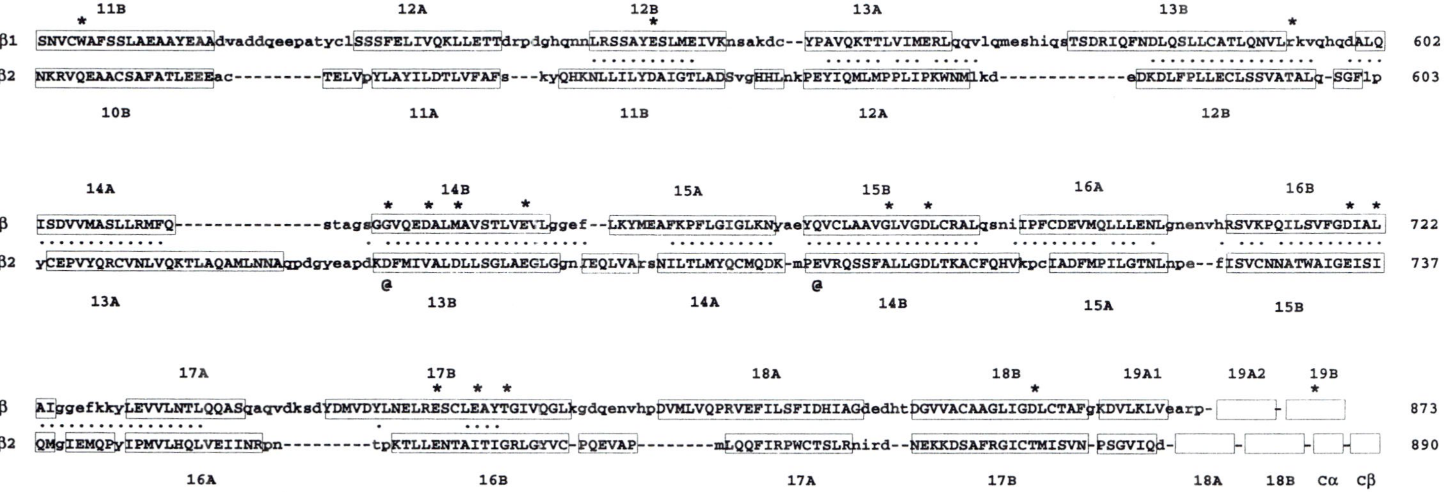

Fig. 7. Structure-based sequence alignment of the two import receptors β (importin-β) and β2 (transportin). Residues in a helical conformation are in upper case and are enclosed in *boxes*. The numbering of the α-helical secondary structure elements is as described in Cingolani et al. (1999), and in Chook and Blobel (1999), respectively. The *dots* indicate residues whose Cα atoms lie within 3.5 Å after optimal superposition of the N-terminal and C-terminal domains as obtained with the program DALI. Residues of β and β2 that interact with Ran are indicated with the symbol @. Residues of β that interact with the IBB cargo are indicated with the symbol * and those that interact with FxFG repeats are indicated with the symbol %. The interactions are as described in the corresponding structure papers (Fig. 2)

In β, two acidic residues in helix 7B (Glu 281 and Asp 288) and the tryptophan residue in the acidic loop (Trp 342) not only interact with RanGTP, but also interact with an essential residue (Arg-13) of the IBB domain. Binding of the IBB cargo and of RanGTP to β are mutually exclusive, as the two molecules would overlap in the central region of β; moreover, they utilize the same key residues for some of their crucial interactions (Fig. 7). The acidic loop is probably also critical for cargo release in β2, although the molecular mechanism of this process is unknown at present. Biochemical studies have mapped the binding of the M9 cargo between HEAT 11 and the C-terminus of β2. Albeit with a different specificity, it is expected that the basic M9 domain will bind via the B helices of the C-terminal arch of β2, where in fact acidic residues are located at structurally identical positions as the acidic residues of β that interact with the IBB (at helices 7B, 12B, 14B, 16B and 17B in Fig. 7).

The prediction from the available structures is that the overall tertiary conformation observed in the two full-length receptors reflect differences both in sequence and in binding partners (Gamblin and Smerdon 1999). It is unlikely that full-length β bound to Ran will assume the conformation observed in the β2-Ran structure. The conformation of the central hinge region between the N-terminal and C-terminal arches is very similar when the IBB and Ran complexes with β are compared, while is quite distinct when β and β2 in the Ran-bound state are compared (Fig. 8). This suggests a predominantly sequence-dependent effect rather than a ligand-dependent effect on the conformation of the central hinge region, which is also the most rigid part of the structures. The central hinge region shows major differences in sequence and topology, with β featuring a longer 7B helix (protruding 20Å from the surface of the molecule) and a much shorter acidic loop (11 residues as compared to the 62 residues in β2). It is also unlikely that full-length β bound to Ran will assume the conformation observed in the β-IBB structure. Ran would in fact sterically clash with the intrarepeat turns of HEAT repeats 13 and 14 of β in the cargo-bound conformation. Nevertheless, in this closed conformation, helix 12B is in the same location as the central portion of the acidic loop of β2, which interacts with Ran in a region that is not contacted in the structure with the β fragment. While a deformation of the superhelix upon RanGTP binding is necessary, the extent of the conformational change required is unclear.

5 Nuclear Pore Complex Docking

RanGTP is instrumental in promoting the final step of a nuclear import process, not only by releasing the cargo but also by releasing the import complex from its association with the NPC (Rexach and Blobel 1995; Görlich et al. 1996b). Importin-/karyopherin-β docks to the NPC by binding to several nucleoporins, which typically contain tandem repeats with a FxFG (Phe-x-Phe-Gly) consensus sequence (see Chaps. 7 and 8). The association of the receptor with the nucleoporin Nsp1 has been analyzed, with the structure of the N-

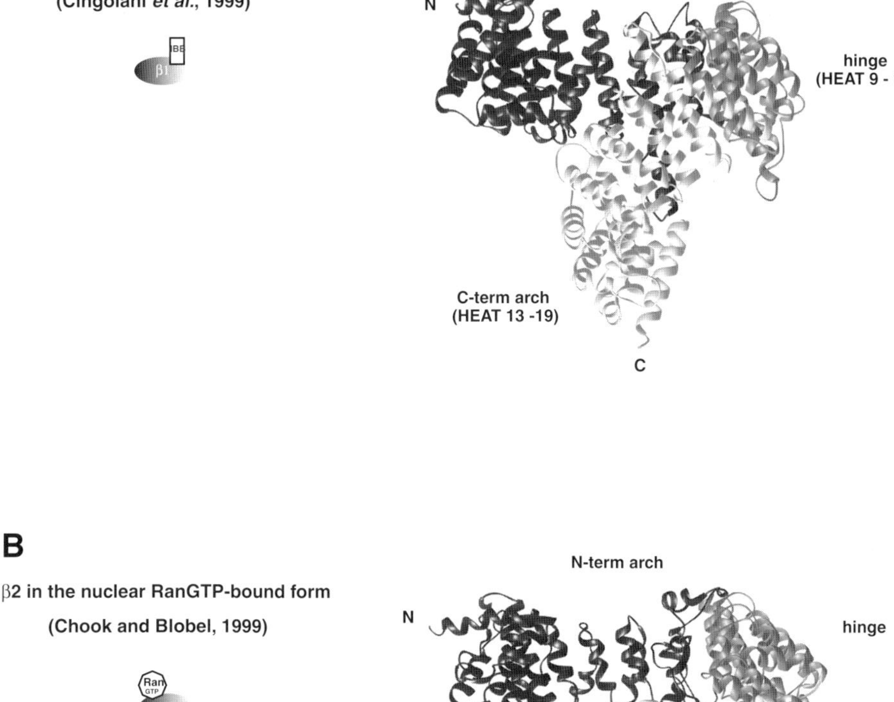

Fig. 8A,B. Conformation of full-length β receptors: **A** β1 when in the IBB complex, and **B** β2 when in the RanGTP complex. The N-terminal arch is in *black*, the hinge region in *dark gray* and the C-terminal arch in *light gray*. The structures have been rotated 90° (with respect to the orientation in Figs. 5A and 6B) along the horizontal superhelical axis of the N-terminal arch

terminal half of β (comprising the first 10 HEAT motifs) in complex with a peptide containing five contiguous FxFG repeats of Nsp1 (Bayliss et al. 2000). β associates with the nucleoporin repeats using its outer convex surface, rather than its inner concave surface as in the case of both cargo and RanGTP binding. The FxFG repeats bind β at helices 5A and 6A, and also at helices 6A and 7A (Fig. 7). The relevance of the hydrophobic interactions observed in the structure is reinforced by the deleterious effects of in vitro binding and in vivo nuclear import caused by mutation of a key hydrophobic residue (Ile 178; Bayliss et al. 2000). The conformation of β in the FxFG and cargo complexes is more similar than in the RanGTP complex. In particular, a relatively small but significant movement of HEAT 5 and 6 occurs in the RanGTP complex and results in the occlusion of the FxFG binding site, providing a mechanism for RanGTP-dependent NPC dissociation (Bayliss et al. 2000). It is puzzling in this context how β-RanGTP docks back to the NPC to be exported into the cytoplasm.

6 From Static Snapshots to a Dynamic Picture

The crystal structures of nuclear import components determined to date provide snapshots of the molecules at different stages in the nuclear import cycle. Given the ensemble of these static images, it is possible to envisage the molecular mechanisms and dynamics of the consecutive series of events that lead to the import of an NLS-containing protein into the nucleus. In the cytosol, the signal sequence of a nuclear protein binds to the NLS-binding domain of importin-/karyopherin-α with precise chemical recognition of crucial positively-charged residues. The IBB domain of α is not involved in NLS binding, and is in fact autoinhibitory due to the internal binding of its C-terminal portion to the α NLS-binding pockets. Both in the presence and in the absence of an NLS-containing protein, the large N-terminal portion of the IBB domain of α is flexible and available to interact with the C-terminal arch of importin-/karyopherin-β. The key initial interaction between α and β is probably the chemical recognition of a few basic residues at the N-terminal portion of the IBB domain with the central hinge region and acidic loop of β, followed by the spiraling of the acidic C-terminal arch of the receptor around the basic IBB helix. The grip around the helix is tightened by the specific interaction of its C-terminal positively charged residue with the C-terminal HEAT repeat of β. The snail-like conformation of β in the cargo-bound complex leaves the N-terminal arch of the receptor open to interact with RanGTP once at the nucleoplasmic side of the NPC (Fig. 5A).

A shape-recognition mechanism rather than precise chemical recognition is responsible for the discrimination of receptors between RanGTP and RanGDP. RanGTP is likely to approach the N-terminal arch of the receptor by docking the switch I region to the first HEAT repeat, and the switch II region more extensively to the contiguous repeat. The binding of the basic patch of Ran to the central hinge region and acidic loop of β would displace the N-terminal portion of the IBB domain and cause at least local changes in the conformation of the C-terminal arch to avoid steric clashes. It is possible to imagine how these concomitant events would progressively loosen the interaction with the IBB domain, which is then dissociated. While this explains the release of an import cargo in the nucleus in case of its direct binding to β, it is unclear how the NLS-containing protein is unloaded from α. Possible mechanisms involve the autoinhibitory effect of the free IBB domain and/or the action of other binding partners of α in the nucleus. Binding of RanGTP also causes α slight change in the conformation of the outer surface of the receptor, hiding the site of Nsp1 nucleoporin binding. The RanGTP-β complex is then recycled to the cytoplasm where the nucleotide is hydrolyzed to GDP. RanGTP is resistant to RanGAP-mediated hydrolysis when bound to the importin, probably due to the inaccessibility of the switch regions when in the complex. RanGTP becomes susceptible to RanGAP-mediated hydrolysis upon binding to the RanBDs of RanBP2 at the cytoplasmic fibril of the NPC or to RanBP1 in the cytoplasm. RanBD interacts with RanGTP using parts of the structure that are exposed to solvent and therefore accessible (Fig. 6 C), in particular the C-terminal tail (Vetter et al. 1999b). As in the case of the importins, the recognition of different regions on the surface of Ran by the receptors and the RanBD proteins allows the consecutive binding of different molecules and therefore progression of the nuclear import cycle.

Although many questions have been answered, many still remain open. How is a non-classical NLS bound to the C-terminus of α in a homologue-specific manner such as in the case of the Stat transcription factor? How is the structure of a native NLS-containing protein modified upon binding to the receptor if a significant stretch of amino acids assumes a totally extended conformation? How drastically is the conformation of β affected when free or when bound to RanGTP? What is the mechanism of cargo binding to other β-like receptors, such as β2? What is the basis for the different binding affinities displayed by different nucleoporins, most of which contain similar FxFG repeats? And last but not least: why does RanGTP binding to exportins result in the association with their cognate cargo in the nucleus rather than in cargo dissociation as observed in the case of the importins? More crystallographic snapshots will be required to fully understand the molecular basis of directionality and specificity of nuclear import and export processes.

Addendum

Recently, the structure of RCC1 bound to Ran has been reported (Renault et al., 2001).

Acknowledgments. I thank Ian Mattaj, Maarten Fornerod and the members of my lab for critical reading of the manuscript.

References

Andrade M, Bork P (1995) HEAT repeats in the Huntington's disease protein. Nat Genet 11: 115–116

Bayliss R, Littlewood T, Stewart M (2000) Structural basis for the interaction between FxFG nucleoporin repeats and importin β in nuclear trafficking. Cell 102:99–108

Chelsky D, Ralph R, Jonak G (1989) Sequence requirements for synthetic peptide-mediated translocation to the nucleus. Mol Cell Biol 9:2487–2492

Chook YM, Blobel G (1999) Structure of the nuclear transport complex karyopherin-β2-GppNHp. Nature 399:230–237

Cingolani G, Petosa C, Weis K, Müller CW (1999) Structure of importin-β bound to the IBB domain of importin-α. Nature 399:221–229

Colledge WH, Richardson WD, Edge MD, Smith AE (1986) Extensive mutagenesis of the nuclear location signal of simian virus 40 large-T antigen. Mol Cell Biol 6:4136–4139

Conti E, Kuriyan J (2000) Crystallographic analysis of the specific yet versatile recognition of distinct nuclear localization signals by karyopherin α. Structure 8:329–338

Conti E, Uy M, Leighton L, Blobel G, Kuriyan J (1998) Crystallographic analysis of the recognition of a nuclear localization signal by the nuclear import factor karyopherin α. Cell 94:193–204

Dang CV, Lee WMF (1988) Identification of the human *c-myc* protein nuclear translocation signal. Mol Cell Biol 8:4048–4054

Dingwall C, Laskey RA (1991) Nuclear targeting sequences – a consensus? Trends Biol Sci 16:178–181

Dingwall C, Sharnick SV, Laskey RA (1982) A polypeptide domain that specifies migration into the nucleus. Cell 30:449–458

Dingwall C, Robbins J, Dilworth SM, Roberts B, Richardson WD (1988) The nucleoplasmin nuclear location sequence is larger and more complex than that of SV40 large T antigen. J Cell Biol 107:841–849

Fontes MR, Teh T, Kobe B (2000) Structural basis of recognition of monopartite and bipartite nuclear localization sequence in mammalian importin α. J Mol Biol 297:1183–1194

Gamblin SJ, Smerdon SJ (1999) Nuclear transport: what a kary-on! Structure 7:R199-R204

Görlich D, Henklein P, Laskey RA, Hartmann E (1996a) A 41 amino acid motif in importin-α confers binding to importin-β and hence transit into the nucleus. EMBO J 15:1810–1817

Görlich D, Kutay U (1999) Transport between the cell nucleus and the cytoplasm. Annu Rev Cell Dev Biol 15:607–660

Görlich D, Pante N, Kutay U, Aebi U, Bischoff FR (1996b) Identification of different roles for RanGDP, RanGTP in nuclear protein import. EMBO J 15:5584–5594

Groves MR, Hanlon N, Turowski P, Hemmings BA, Barford D (1999) The structure of the protein phosphatase 2A PR65/A subunit reveals the conformation of its 15 tandemly repeated HEAT motifs. Cell 96:99–110

Henderson BR, Percipalle P (1997) Interactions between HIV Rev and nuclear import and export factors: the Rev nuclear localisation signal mediates specific binding to human importin-β. J Mol Biol 274:693–707

Hillig RC, Renault L, Vetter IR, Drell TT, Wittinghofer A, Becker J (1999) The crystal structure of Rna1p: a new fold for a GTPase-activating protein. Mol Cell 3:781–791

Jacobs MD, Harrison SC (1998) Structure of an IκBα/NF-κB complex. Cell 95:749–758

Kalderon D, Roberts BL, Richardson WD, Smith AE (1984) A short amino acid sequence able to specify nuclear location. Cell 39:499–509

Kobe B (1999) Autoinhibition by an internal nuclear localization signal revealed by the crystal structure of mammalian importin α. Nat Struct Biol 6:388–397

Kutay U, Izaurralde E, Bischoff FR, Mattaj IW, Görlich D (1997) Dominant-negative mutants of importin-β block multiple pathways of import and export through the nuclear pore complex. EMBO J 16:1153–1163

Lanford RE, Butel JS (1984) Construction and characterization of an SV40 mutant defective in nuclear transport of T antigen. Cell 37:801–813

Macara IG (1999) Nuclear transport: randy couples. Curr Biol 9:R436-R439

Makkerh JPS, Dingwall C, Laskey RA (1996) Comparative mutagenesis of nuclear localization signals reveals the importance of neutral and acidic amino acids. Curr Biol 6:1025–1027

Malik HS, Eickbush TH, Goldfarβ DS (1997) Evolutionary specialization of the nuclear targeting apparatus. Proc Natl Acad Sci USA 94:13738–13742

Mattaj IW, Englmeier L (1998) Nucleocytoplasmic transport: the soluble phase. Annu Rev Biochem 67:265–306

Moroianu J, Blobel G, Radu, A (1996) The binding site of karyopherin alpha for karyopherin β overlaps with a nuclear localization sequence. Proc Natl Acad Sci USA 93:6572–6576

Müller CW, Rey FA, Sodeoka M, Verdine GL, Harrison SC (1995) Structure of the NF-κB p50 homodimer bound to DNA. Nature 373:311–317

Nakielny S, Dreyfuss G (1999) Transport of proteins and RNAs in and out of the nucleus. Cell 99:677–690

Peifer M, Berg S, Reynolds AB (1994) A repeating amino acid motif shared by proteins with diverse cellular roles. Cell 76:789–791

Renault L, Nassar N, Vetter I, Becker J, Klebe C, Roth M, Wittinghofer A (1998) The 1.7Å crystal structure of the regulator of chromosome condensation (RCC1) reveals a seven-bladed propeller. Nature 392:97–101

Renault L, Kuhlmann J, Henkel A, Wittinghofer A. (2001) Structural basis for guanine nucleotide exchange on Ran by the regulator of chromosome condensation (RCC1). Cell 105:245–55

Rexach M, Blobel G (1995) Protein import into nuclei: association and dissociation reactions involving transport substrate, transport factors, and nucleoporins. Cell 83:683–692

Robbins J, Dilworth SM, Laskey RA, Dingwall C (1991) Two interdependent basic domains in nucleoplasmin targeting sequence: identification of a class of bipartite nuclear targeting sequence. Cell 64:615–623

Scheffzek K, Klebe C, Fritz-Woolf K, Kabsch W, Wittinghofer A (1995) Crystal structure of the nuclear Ras-related protein Ran in its GDP-bound form. Nature 374:378–381

Stewart M, Kent HM, McCoy AJ (1998) Structural basis for molecular recognition between nuclear transport factor 2 (NTF2) and the GDP-bound form of the Ras-family GTPase Ran. J Mol Biol 277:635–646

Truant R, Cullen BR (1999) The arginine-rich domains present in human immunodeficiency virus type I Tat and Rev functions as direct importin β-dependent nuclear localization signals. Mol Cell Biol 19:1210–1217

Vetter I R, Arndt A, Kutay U, Görlich D, Wittinghofer A (1999a) Structural view of the Ran-importin β interaction at 2.3 Å resolution. Cell 97:635–646

Vetter I R, Nowak C, Nishimoto T, Kuhlmann J, Wittinghofer A (1999b) Structure of a Ran-binding domain complexed with Ran bound to a GTP analogue: implications for nuclear transport. Nature 396:474–477

Weis K (1998) Importins and exportins: how to get in and out of the nucleus. Trends Biochem Sci 23:185–189

Weis K, Ryder U, Lamond AI (1996) The conserved amino-terminal domain of hSRP1 α is essential for nuclear protein import. EMBO J 15:7120–7128

Nuclear Export of tRNA

George Simos[1]*, Helge Großhans[1], and Ed Hurt[1]

1 Introduction

In the past few years, the study of transport of macromolecules between the nucleus and the cytoplasm has produced an overwhelming amount of data which have been formulated into convincing mechanistic models of nucleocytoplasmic transport. We will briefly introduce these current models in the field of nucleocytoplasmic transport (recently reviewed in Adam 1999; Görlich and Kutay 1999; Nakielny and Dreyfuss 1999; Sträßer and Hurt 1999; Talcott and Moore 1999; Ryan and Wente 2000; Wente 2000, and also treated in the other chapters of this volume) before focusing on the nuclear export of one particular type of transport substrate (cargo), namely transfer RNA (tRNA), which is synthesized and processed in the nucleus, but required for protein biosynthesis in the cytoplasm (for other recent reviews on tRNA transport see Simos 1999; Wolin and Matera 1999; Großhans et al. 2000b). Nuclear import as well as nuclear export of proteins and RNA occurs through the nuclear pore complexes (NPC) and is usually an active, carrier-mediated process. Proteins that are destined to enter the nucleus are recognized and bound in the cytoplasm by receptors (named importins or karyopherins) that mediate targeting and translocation through the NPC. Inside the nucleus, the receptor releases its import cargo and is then recycled to the cytoplasm. Importins can either associate directly with their transport cargo recognizing a nuclear localization sequence or require adaptor molecules. All the importins identified so far, are members of the same protein family (named importin β family after its founding member) and contain a characteristic, conserved Ran-GTP binding domain.

The general principles of nuclear protein import also apply to export of proteins and some RNA species from the nucleus. Importin β family members have been shown to be involved in nuclear export processes and were subsequently termed exportins. One of these, CRM1, is the export receptor for the leucine-rich nuclear export signal (NES) found in many different proteins.

[1] Biochemie-Zentrum Heidelberg (BZH), Im Neuenheimer Feld 328, 69120 Heidelberg, Germany

* *Present address:* G. Simos, Laboratory of Biochemistry, School of Medicine, University of Thessaly, 22 Papakiriazi Str., 41222 Rarissa, Greece

Results and Problems in Cell Differentiation, Vol. 35
K. Weis (Ed.): Nuclear Transport

Thus, it is conceivable that, similar to the adaptors used in protein import, NES-containing proteins that can bind to RNA can serve as adaptors in RNA nuclear export. This appears indeed to be true for at least a subset of RNAs, such as viral mRNAs and snRNAs. The small GTPase Ran and its effectors play a central role in nucleocytoplasmic transport (Azuma and Dasso 2000). According to the current view, nuclear Ran is in the GTP-bound form (Ran-GTP) whereas in the cytoplasm Ran-GDP predominates. This is because of the exclusive nuclear localization of the Ran nucleotide exchange factor RCC1 (Prp20p in yeast) and the cytoplasmic localization of the GTPase activating protein RanGAP1 (Rna1p in yeast). Binding of Ran-GTP to an importin can trigger the dissociation of the importin/import substrate complex, whereas, on the other hand, binding to an exportin promotes the association with the corresponding export cargo and the formation of an export competent complex. Hydrolysis of the Ran-bound GTP, which should occur only in the cytoplasm, can then cause the dissociation of Ran from the exportin and the release of the export substrate.

Despite its simplicity and apparent universality, the model described above, seems not to be directly applicable to all nucleocytoplasmic transport processes. There have been several reports on proteins that can enter the nucleus without the help of importins (Görlich and Kutay 1999; Hetzer and Mattaj 2000). These proteins may have acquired the capacity to directly associate with components of the NPC in a way that mimics the interaction between importins and NPC, and therefore triggers their translocation across the nuclear membrane. Two major transport processes, namely nuclear export of mRNA and ribosomal subunits, appear not to be mediated by any of the known exportins either, and the involvement of Ran in these processes is still controversial (Görlich and Kutay 1999; Sträßer and Hurt 1999). Both mRNA and rRNA leave the nucleus in association with many proteins as large ribonucleoprotein particles (RNPs) or pre-ribosomal subunits, respectively. Nuclear export of these particles is clearly more complex than transport of single proteins, and this may not simply be because of the size differences between these two types of cargo. The passage of an RNA through the NPC is likely to represent an integral part of its maturation process and as such it may strongly depend on upstream, i.e., nuclear, and downstream, i.e., cytoplasmic, RNA processing events. Nuclear export of the second most abundant class of RNA, tRNA, does indeed exhibit such characteristics as it is apparently coupled to other tRNA biogenesis steps such as modification, splicing, or aminoacylation. However, unlike mRNA or rRNA, tRNA species, probably because of their small size and uniform three-dimensional structure, are also capable of actively exiting the nucleus simply by associating directly with a member of the importin β family, which acts as a specialized exportin for tRNA. Before describing this mechanism in detail, we will briefly discuss the processing of tRNA and refer to the early findings that paved the road for the study of tRNA transport.

2 tRNA Processing: Preparation to Exit the Nucleus?

All tRNA molecules are synthesized by RNA polymerase III as precursors that undergo a complex maturation process, before they are exported from the nucleus. These events include trimming of the 5′ and 3′ ends, addition of three terminal CCA residues, modification of a number of nucleosides and, in the case of some tRNAs, splicing (reviewed in Hopper and Martin 1992; Westaway and Abelson 1995; Wolin and Matera 1999). It was initially thought that removal of introns from pre-tRNAs occurs after 5′- and 3′-end processing, making splicing the last tRNA maturation step prior to export through the nuclear pores. This idea was supported by microinjection studies in *Xenopus* oocytes which showed that end processing preceded intron removal (Melton et al. 1980) and was confirmed with the detection of end-matured but intron-containing tRNA processing intermediates in yeast (Hopper and Martin 1992). The observation that in yeast the key enzymes of splicing, tRNA endonuclease and tRNA ligase, were found associated with the inner side of the nuclear membrane, raised then the interesting possibility that the tRNA-splicing reaction might be coupled to the translocation step through the nuclear pores (Culbertson and Winey 1989; Westaway and Abelson 1995). In agreement with this idea, several yeast nucleoporin mutants were shown to be defective in pre-tRNA splicing (Sharma et al. 1996).

However, subsequent studies in yeast showed that splicing and end trimming are not necessarily ordered relative to each other because it was possible to detect processing intermediates that were spliced but contained 5′ and 3′ extensions (O'Connor and Peebles 1991). Finally, it has recently been shown that the processing intermediates that accumulate in *Xenopus* oocyte nuclei after injection of tRNA transcripts are spliced but not end-matured (Lund and Dahlberg 1998). Only when high amounts of precursor were injected, end-matured but intron-containing intermediates accumulated, and this might have been the case in the earlier microinjection studies. These later data suggest that under physiological conditions, i.e., when only low amounts of pre-tRNAs are found in the nucleus, splicing is kinetically favored and occurs prior to removal of terminal sequences (Lund and Dahlberg 1998). The tRNA splicing machinery, however, seems to be more easily saturated when an excess of precursor tRNA is present in the nucleus, leading to the accumulation of unspliced, but otherwise mature, intermediates.

According to these recent results, the cargo that is transported from the nucleus into the cytoplasm should be both end-trimmed and spliced tRNAs. But can tRNA molecules which have not undergone maturation leave the nucleus? Both earlier and recent microinjection studies have shown that unspliced tRNAs can enter the cytoplasm, when produced in excess (Melton et al. 1980; Lund and Dahlberg 1998). On the other hand, tRNAs that contain unprocessed ends, are always confined in the nucleus. This suggests that mature ends, including the CCA 3′ terminal sequence, are a pre-requisite for nuclear tRNA export while the presence of introns can be tolerated.

3 A Distinct Pathway for the Nuclear Export of Mature tRNA

tRNA was one of the first nucleocytoplasmic transport substrates to be studied in *Xenopus* oocytes, and its export was shown to be a temperature-dependent and saturable, therefore carrier-mediated process (Zasloff 1983). Furthermore, it was reported that different tRNA species shared, at least in this system, a common carrier system. Transport of a particular tRNA species, human initiator $tRNA^{Met}$, was shown to be sensitive to mutations that mapped in the highly conserved D stem-loop and T stem-loop regions of the tRNA. Strikingly, all mutations that impaired transport also caused a tRNA processing defect, suggesting that the tRNA transport mechanism has similar structural requirements as the processing enzymes, including correct tertiary folding and stabilization of the L-shape of tRNA (Tobian et al. 1985). The transport of tRNA through the NPCs has also been directly visualized by coupling tRNA molecules to colloidal gold particles injected into *Xenopus* nuclei. These particles were shown to sequentially associate with the nuclear substructure of the NPCs, to cross the central gated channel and to be released into the cytoplasm (Dworetzky and Feldherr 1988; Panté et al. 1997). Dissociation of the tRNA from intranuclear retention sites also proved to be a determinant for export as it may represent a rate-limiting step for the transport reaction (Pokrywka and Goldfarb 1995). In fact, nuclear retention, and not lack of interaction with the export machinery, was shown to be the cause for the inability of a particular mutant tRNA to exit the nucleus (Boelens et al. 1995)

A key finding in the RNA export field, has been the demonstration that the export of different classes of RNA (tRNA, 5S rRNA, U snRNA, mRNA) microinjected into *Xenopus* oocyte nuclei is mediated by different saturable factors, as the export of a given class of RNA can be saturated by itself but not by other RNAs, e.g., nuclear export of tRNA is not efficiently competed by mRNA, rRNA and snRNA (Jarmolowski et al. 1994). It thus appears that different classes of RNA depend for their export on specific signals, which could either lie in the RNA moiety or be provided by proteins with which the RNAs associate. tRNA nuclear export displayed some additional specific features when compared with other RNA species such as faster kinetics and insensitivity to competition by homopolymeric RNAs that inhibited the export of 5 S RNA, mRNA, and U snRNAs. Furthermore, nuclear export of tRNA was not affected by antibodies against the nucleoporin Nup98 or by the matrix protein of vesicular stomatitis virus, which impaired export of most other cellular RNAs (Her et al. 1997; Powers et al. 1997).

The involvement of known transport factors such as Ran in the nuclear export of tRNA was originally a controversial issue. It was thought at first that tRNA nuclear export may be independent of Ran because it was not affected in mammalian cells carrying a mutant RCC1, in which nuclear export of mRNA and snRNA was inhibited (Cheng et al. 1995). However, it was subsequently shown that nuclear export of tRNA from *Xenopus* oocyte nuclei required the presence of Ran-GTP in the nucleus, as its depletion by nuclear injection of RanGAP blocked tRNA export (Izaurralde et al. 1997). Furthermore, the nuclear export of tRNA from *Xenopus* oocyte nuclei could be significantly impaired by the injection of dominant-negative importin β mutants (Kutay et al. 1997). The involvement of Ran in tRNA export had also been suggested by findings in yeast. The *rna1-1* mutant was shown to be defective in pre-tRNA splicing as well as processing of pre-rRNA and production and export of mRNA from the nucleus (Hopper et al. 1978; Amberg et al. 1992). It was subsequently discovered that Rna1p is a cytoplasmic protein (Hopper et al. 1990) that corresponds to the yeast RanGAP (Becker et al. 1995). Thus, the accumulation of unspliced pre-tRNAs in the *rna1-1* mutant appears to be caused by a defect in the nucleocytoplasmic transport machinery. This might be an indirect effect, e.g., caused by defective nuclear import of tRNA-processing or export factors, or might reflect a direct requirement of an active Ran cycle for tRNA nuclear export. In either case, the involvement of Ran in tRNA export has been directly demonstrated by the localization of tRNAs in yeast using fluorescence in situ hybridization. Yeast cells carrying a mutant thermosensitive Rna1p were shown to accumulate both mature and unspliced tRNAs in the nucleus when shifted to the non-permissive temperature (Sarkar and Hopper 1998; Großhans et al. 2000a).

4 Identification of Los1p/Xpo-t as a Nuclear Export Receptor for tRNA

Initial evidence for a role of the yeast protein Los1p in nuclear tRNA export was obtained several years ago. The non-essential *LOS1* gene had been identified in a screen for mutants exhibiting conditional loss of activity of a suppressor-tRNA^{Tyr} (LOS standing for loss of suppression; Hopper et al. 1980). *los1* mutants produced reduced levels of functional suppressor tRNA and accumulated end-trimmed but unspliced pre-tRNA, suggesting a function of Los1p in the splicing of tRNA. However, cells that lacked Los1p contained apparently normal levels of tRNA splicing endonuclease and ligase activities (Hurt et al. 1987; Shen et al. 1993, 1996). A role for Los1p in the transport of tRNA became more likely when it was shown that Los1p not only interacts genetically with the essential nucleoporin Nsp1p but also physically associates with it and localizes to nuclear pore complexes (Simos et al. 1996b). The idea that Los1p may be a nuclear pore associated factor required for the nuclear export step of tRNA biogenesis was further strengthened when sequence homology searches showed that Los1p contains an amino-terminal Ran-GTP-binding motif that characterizes the importin β protein family (Görlich et al. 1997).

Indeed, Los1p was shown experimentally to meet major requirements for a tRNA exportin; i.e., it binds to Ran-GTP as well as to two repeat containing nuclear pore proteins (Nsp1p and Nup2p) in vivo, and interacts with tRNA in a Ran-GTP-dependent manner in vitro (Hellmuth et al. 1998). Furthermore, over-expression of Los1p was shown to inhibit other nuclear export pathways most likely by saturating the NPC binding sites involved in the translocation of nuclear export complexes to the cytoplasm (Hellmuth et al. 1998). Taken together, these results strongly suggested that yeast Los1p is a bona fide member of the importin β-like family with a function in nuclear export of tRNA. The final in vivo proof for this idea came from tRNA localization experiments. Using fluorescent in situ hybridization it was shown that yeast cells that lack Los1p exhibit a weak accumulation of mature major $tRNA^{Ile}$ inside the nucleus (Sarkar and Hopper 1998). This export defect became much more evident when tRNA species that undergo splicing, such as $tRNA^{Leu}$ or minor $tRNA^{Ile}$, were examined (Großhans et al. 2000a). Localization of the mature (spliced) forms of these tRNA species revealed strong accumulation inside the nucleus and predominantly the nucleolus. [The nucleolus in general appearing to be the site of preferred tRNA accumulation if nuclear export, for whatever reason, cannot proceed (Großhans et al. 2000a).] However, under the same conditions export of two tRNA species that lack introns, $tRNA^{Glu}$ and $tRNA^{Gly}$, was very little affected. This preferential requirement of Los1p for the efficient nuclear export of the tRNAs encoded by intron-containing genes may, in agreement with the earlier findings discussed above, reflect an additional role of Los1p in intranuclear transport steps that splicing necessitates. One possibility is that Los1p binds to intron-containing pre-tRNAs and facilitates their processing by presenting them to the tRNA splicing machinery, before taking them out of the nucleus. It cannot be ruled out, however, that the splicing defect observed in the absence of Los1p, is a secondary effect of intra-nuclear tRNA accumulation, mediated by an unknown feedback mechanism that inhibits the splicing machinery.

The human homologue of Los1p (termed Exportin-t or Xpo-t) was first identified by its homology to Los1p (21% identity; Arts et al. 1998a) or as a protein from Hela extracts that bound to Ran-GTP (Kutay et al. 1998). Xpo-t was shown to shuttle between the nucleus and the cytoplasm and to bind cooperatively to Ran-GTP and tRNA in vitro. Binding was stronger to native tRNAs than to in vitro transcripts, suggesting that modifications are important for this interaction (Kutay et al. 1998). When Xpo-t was injected into the nuclei of *Xenopus* oocytes together with labeled tRNAs, it stimulated their nuclear export (Arts et al. 1998a; Kutay et al. 1998). Furthermore, injection of antibodies against Xpo-t into the oocyte nuclei significantly inhibited tRNA nuclear export suggesting that Xpo-t is the major, if not the only, tRNA export receptor in *Xenopus* oocytes (Arts et al. 1998b; Lipowsky et al. 1999).

Taking into account all the data discussed above, the following model for Los1p/Xpo-t mediated nuclear export of tRNA was suggested. In the nucleus, Los1p/Xpo-t interacts with Ran-GTP and newly synthesized mature tRNAs forming a ternary complex. This complex is targeted to the nuclear pores through the affinity of Los1p/Xpo-t for nucleoporins such as, in the case of yeast, Nsp1p and Nup2p. Nup2p has been localized to the nuclear side of the

NPC and has been suggested to facilitate the export of Srp1p (the yeast importin α) by the exportin Cse1p (Booth et al. 1999; Hood et al. 2000). Because Nup2p contains a Ran-binding domain and can interact in vivo with Los1p, it may also be involved in the stabilization of the Los1p/Ran-GTP/tRNA complex and its initial docking to the NPC. Nsp1p, on the other hand, which is both genetically and physically linked to Los1p, has been localized on both sides of the central gated channel of the NPC (Fahrenkrog et al. 1998; Rout et al. 2000), so it may be involved in translocation of the tRNA export complex through the central part of the NPC, a step about which very little is known. Once the ternary complex is on the cytoplasmic side, hydrolysis of GTP to GDP can be stimulated by the concerted action of RanGAP (Rna1p in yeast) and RanBP1 (Yrb1p in yeast) so that the export complex dissociates. Los1p/Xpo-t as well as Ran-GDP are re-cycled into the nucleus while tRNA is delivered to the translation machinery. The direct interaction of tRNA with an exportin is probably the main difference from other RNA classes, which so far have been shown to depend on adaptor RNA-binding proteins in order to gain access to the export machinery. This could explain the specific characteristics of the tRNA nuclear export pathway reported earlier, such as the fast export kinetics or the insensitivity to reagents that block export of other RNAs.

Apart from Los1p/Xpo-t, a few other proteins with diverse functions have been implicated in tRNA export mainly because of their capacity to interact with tRNA. The glycolytic enzyme GAPDH (glyceraldehyde-3-phosphate dehydrogenase) was shown to bind to nuclear export competent tRNAs but not to mutants defective in export (Singh and Green 1993). However, GAPDH was subsequently shown to bind with higher affinity to AU-rich mRNA (Nagy and Rigby 1995). Yeast zuotin, initially identified as a nuclear protein that binds to left-handed Z-DNA (Zhang et al. 1992) was also shown to associate with tRNA (Wilhelm et al. 1994). However, zuotin was then reported to be located predominantly in the cytoplasm and to associate with ribosomes, probably acting in a DnaJ-like fashion as a chaperone for the nascent polypeptide chain (Yan et al. 1998). Finally, human vigilin, a multiple KH-domain protein, was reported to bind to tRNA, to stimulate weakly its nuclear export and to be part of a multiprotein complex that also contained Xpo-t (Kruse et al. 2000). On the other hand, the *Xenopus* vigilin was shown to bind to the 3′ UTR of vitellogenin mRNA (Dodson and Shapiro 1997) while the *Drosophila* homologue was found to interact with centromeric heterochromatin (Cortes et al. 2000). The yeast homologue of vigilin, Scp160p, has been demonstrated to be associated with polyribosomes, probably as a component of mRNP complexes (Weber et al. 1997; Lang and Fridovich-Keil 2000). Overall, the data on roles of GAPDH, zuotin or vigilin in nuclear tRNA export are not consistent and their potential involvement in this process has yet to be convincingly demonstrated.

5 Structural Requirements for the Xpo-t/tRNA Interaction: A Proofreading Mechanism?

Xpo-t/Los1p is so far unique among the members of the karyopherin family because it can interact directly with RNA. Chemical and enzymatic cleavage/protection and binding interference experiments allowed a first structural analysis of this interaction (Arts et al. 1998b). Furthermore, the in vitro reconstitution of the Xpo-t/tRNA/Ran-GTP complex allowed the comparison of the export rates of various mutant tRNAs as determined in the *Xenopus* system and their respective affinities to Xpo-t (Arts et al. 1998b; Lipowsky et al. 1999). The foot-printing experiments suggested that the protein–tRNA interface in the Xpo-t/Ran-GTP/tRNA complex consists of the acceptor and TΨ C arms, which are then likely to be the sites recognized by Xpo-t. However, a minihelix consisting of these two elements did not suffice for binding to Xpo-t, suggesting that additional structural features are recognized. Indeed, analysis of mutant tRNAs for binding in vitro and export in vivo demonstrated a requirement for mature 5′ and 3′ ends, including the addition of the CCA nucleotides at the 3′ end. On the other hand, an intron containing tRNA bound equally well to Xpo-t/Ran-GTP as a spliced tRNA. Nevertheless, the nuclear export of an unspliced tRNA was poor unless an excess of Xpo-t was co-injected in the oocyte nuclei suggesting that, normally, unspliced tRNAs are retained in the nucleus through their interaction with other intranuclear proteins.

Mutations that disturb the secondary structure of the T stem (such as in position 52) or the tertiary base pairing between the T and the D loop (e.g., the C56-G19 base pair) were already reported in the early studies to prevent nuclear tRNA export (Tobian et al. 1985). These mutations were shown to abolish also binding of tRNA to Xpo-t (Arts et al. 1998b; Lipowsky et al. 1999). Another factor that influences the affinity of a tRNA for Xpo-t was shown to be its modification state (Lipowsky et al. 1999). A native, i.e., fully modified, tRNA was shown to bind much stronger to Xpo-t than the corresponding unmodified in vitro transcript. Finally, Xpo-t was shown to interact equally well with tRNAs of both pro-and eukaryotic origin, confirming the fact that it recognizes universal tRNA structural features (Lipowsky et al. 1999).

In summary, correct tRNA shape, properly matured 5′ and 3′ ends and base modifications are all positive determinants for efficient in vitro binding to Xpo-t. The fact that the same elements are also required for efficient nuclear export in vivo, supports the notion that Xpo-t could be the main mediator of nuclear tRNA export, at least in *Xenopus* oocytes. Furthermore, binding to Xpo-t may also act as a quality control check-point that would monitor the structural integrity and thus functionality of the tRNAs before they are allowed to exit the nucleus. An exception to this proofreading capacity of Xpo-t is its inability to monitor the presence or absence of an intron. However, although unspliced pre-tRNAs can in principle exit the nucleus through their binding to Xpo-t, they are normally confined to the nucleus. As mentioned previously, this can be explained by the kinetic control of tRNA maturation in *Xenopus* oocytes, i.e., by the fact that splicing normally occurs before end-processing, so that intron-containing tRNAs should have no access to Xpo-t because of their immature ends. In yeast, on the other hand, the ability of unspliced tRNAs to bind to their exportin may have functional relevance, as splicing may be coupled to nuclear export (as discussed above). It should however be noted that all the binding studies performed so far concern the mammalian protein Xpo-t. The results of these studies may not be directly applicable to the yeast homologue Los1p, because of the low level of identity between the two proteins. The possible tRNA sequence elements required for nuclear export in yeast have been investigated in a recent study (Cleary and Mangroo 2000). Various mutants of a suppressor tRNA were analyzed for functionality as well as localization using biochemical fractionation. The results indicate that a mutation in position 11 at the D stem causes nuclear retention, which can be rescued by over expression of Los1p. This type of approach complemented by in vitro binding assays is expected to give, in the future, information about the requirements for an efficient Los1p/tRNA interaction.

6 An Alternative Nuclear tRNA Export Pathway and the Role of Aminoacylation

Although the model for Los1p/Xpo-t-mediated tRNA nuclear export can explain satisfactorily most of the existing data, it may be far from complete. A puzzling issue is the fact that Los1p is not essential for viability in yeast cells as disruption of the *LOS1* gene causes no apparent growth phenotype (Hurt et al. 1987; Simos et al. 1996b). Furthermore, as shown by the tRNA localization studies discussed above, certain tRNA species are very little or not at all affected in their nuclear export when the *LOS1* gene is disrupted (Großhans et al. 2000a). Thus, there must be an alternative, Los1p-independent pathway that allows tRNA export to continue in the absence of Los1p. Interestingly, the Los1p function becomes essential, if other components of the tRNA biogenesis machinery are mutated, such as Tfc4p, a subunit of the tRNA transcription factor TFIIIC, Pus1p, a nuclear tRNA pseudouridine synthase, or Arc1p, a tRNA-binding protein that associates stably with two aminoacyl-tRNA synthetases, MetRS and GluRS (Simos et al. 1996a, 1996b, 1998; Motorin et al. 1998). A possible explanation for the synthetic lethality observed in these cases may be the synergism between the weak nuclear tRNA export defect caused by the absence of Los1p and the reduction in the availability of functional tRNA caused by the mutations in the aforementioned proteins. However, it is equally likely that these proteins may be components of an independent tRNA export system that functions parallel to Los1p. According to recent findings, this system in addition appears to involve aminoacylation of tRNA as well as the essential translation factor eEF-1A (Sarkar et al. 1999; Großhans et al. 2000a). In this context we will briefly describe the function of the proteins that interact genetically with Los1p and discuss the link of nuclear tRNA export to the tRNA aminoacylation and protein translation machineries.

Tfc4p is an essential component of the TFIIIC transcription factor complex that regulates RNA polymerase III-dependent transcription of tRNA genes by binding to intragenic promoter elements and recruiting transcription factor TFIIIB (Marck et al. 1993). A mutation in *TFC4*, which causes a thermosensitive phenotype and reduced tRNA transcription rates, leads to lethality in the absence of Los1p (Simos et al. 1996b). The *PUS1* gene was identified through its genetic relationship with *LOS1* and the nucleoporin gene *NSP1* because a combination of mutations in these three genes caused synthetic lethality (Simos et al. 1996b). Despite the fact that Pus1p was shown to catalyze the formation of pseudouridines in several positions (mainly 27, 28, 34, and 36) and in a number of different tRNA species, disruption of its gene conferred no obvious phenotype, as if all the modifications formed by Pus1p are dispensable for normal cell growth (Motorin et al. 1998). However, when disruption of *PUS1* was combined with mutations in *LOS1*, the cells grew very slowly at 30°C and ceased growth at 37°C. The genetic link between Pus1p, which localizes exclusively inside the nucleus, and Los1p can be explained if Pus1p-catalysed modifications act as positive determinants for association of tRNA with the components of an alternative tRNA nuclear export machinery. This hypothesis is supported by the fact that almost all pseudouridines that are formed by Pus1p are found in positions that are specific for eukaryotic tRNA. Therefore, it is likely that Pus1p has a specific role in a strictly eukaryotic tRNA biogenesis step, such as the nuclear export of tRNA.

Arc1p associates both in vitro and in vivo with two aminoacyl-tRNA synthetases, methionyl-tRNA (MetRS) and glutamyl-tRNA (GluRS) synthetase (Simos et al. 1996a, 1998). This association is mediated by the amino-terminal domain of Arc1p whereas its carboxy terminus harbors a novel tRNA-binding domain (TRBD). Arc1p alone binds to several tRNA species, but when in complex with GluRS and MetRS, it promotes the efficient interaction of the complex only with the cognate tRNAs. Mutant yeast cells that lack the TRBD grow almost as well as wild-type cells. However, when the absence of the TRBD is combined with mutations in Los1p or Nsp1p the cells are no longer able to grow. The TRBD of Arc1p is highly conserved in the protein p43/EMAPII, the non-enzymic component of the mammalian multisynthetase complex, which contains nine aminoacyl-tRNA synthetases including GluRS and MetRS (Quevillon et al. 1997). The TRBD is also found fused to the C-terminus of the human tyrosyl-tRNA synthetase (Kleeman et al. 1997), which is not part of the multisynthetase complex, and to *C. elegans* MetRS (Wilson et al. 1994) suggesting that the TRBD can function both intra- and inter-molecularly. Although Arc1p is found mainly in the cytoplasm, its mammalian homologue, p43, has been localized by immunoelectron microscopy predominantly inside the nucleus (Popenko et al. 1994). The strong genetic link between Los1p and a component of the tRNA aminoacylation machinery hinted at the possibility that aminoacylation of tRNA might be required for the efficient nuclear export of tRNA. This idea appeared paradoxical at that time, because tRNA-aminoacylation, despite the fact that some components of the aminoacylation machinery had previously been detected in the nucleus (Kisselev and Wolfson 1994), was considered to be a strictly cytoplasmic process. Therefore, the observation that tRNAs can also undergo aminoacylation inside the nucleus came rather as a surprise (Lund and Dahlberg 1998).

tRNAs injected in *Xenopus* oocyte nuclei were shown to be charged with their cognate amino acids before their exit from the nucleus or even when export was inhibited (Lund and Dahlberg 1998). Unexpectedly, an unspliced $tRNA^{Tyr}$ could also be aminoacylated in the nucleus. The presence of specific aminoacylation inhibitors not only abolished intranuclear as well as cytoplasmic aminoacylation but also led to significant retardation in the nuclear export of the non-aminoacylated tRNA (Lund and Dahlberg 1998). How aminoacylation acts on nuclear export of tRNA is not clear. The fact that Xpo-t binds non-aminoacylated tRNAs with high affinity (Kutay et al. 1998; Lipowsky et al. 1999) argues against a strong influence of aminoacylation on the association with this particular export receptor. However, it is still unknown whether the affinity of Xpo-t to tRNA might yet be increased by aminoacylation – similar to the effect of base modifications (see above). In any case, the requirement for aminoacylation is not an absolute one in the *Xenopus* system, as mutant tRNAs that cannot be aminoacylated are still capable of exiting the nucleus, albeit with slower kinetics (Arts et al. 1998b; Lipowsky et al. 1999). Therefore, intranuclear aminoacylation may only facilitate nuclear tRNA export probably by releasing mature tRNAs from the processing machinery and, at the same time, proofreading their structural integrity.

Several lines of in vivo evidence demonstrate the importance of aminoacylation for nuclear tRNA export in the yeast system. Starvation of auxotrophic yeast cells for the amino acid leucine, a treatment that reduces the aminoacylation levels of $tRNA^{Leu}$, was shown to cause intranuclear accumulation of this tRNA. Inhibition of isoleucyl-tRNA synthetase by an amino acid analog added to the medium, strongly impaired the nuclear export of the cognate $tRNA^{Ile}$. In both cases the export of other tRNA species remained unaffected (Großhans et al. 2000a). Temperature-sensitive mutants of methionyl-, isoleucyl- or tyrosyl-tRNA synthetases were found to accumulate tRNA inside the nucleus at the restrictive temperature (Sarkar et al. 1999). This effect was specific for the cognate tRNA in the case of the methionyl-tRNA synthetase mutant but in the other two mutants also non-cognate tRNAs were affected. Accumulation of tRNAs inside the nuclei was also observed upon inactivation of Cca1p using the temperature-sensitive *cca1-1* mutant (Sarkar et al. 1999; Großhans et al. 2000a). Cca1p completes the maturation of the 3′ ends of all tRNAs by adding the three terminal CCA bases, which is essential for aminoacylation but may also be required in yeast for recognition of tRNAs by Los1p. Finally, a mutant $tRNA^{Val}$ that is inefficiently aminoacylated was shown to accumulate inside the nucleus when expressed in yeast cells (Qiu et al. 2000).

Taken together, these findings demonstrate the involvement of aminoacylation in nuclear tRNA export in both *Xenopus* oocytes and vegetatively growing yeast cells, but raise a number of intriguing questions. First of all, what is the connection between the aminoacylation-dependent and the Los1p-dependent tRNA export pathways? It is possible that these are two distinct but redundant export routes. Indeed, the absence of Los1p aggravates the nuclear tRNA export defect caused by reduced levels of aminoacylation upon amino acid starvation (Großhans et al. 2000a). It is likewise possible that the requirement for aminoacylation applies to all tRNA species but some of them, i.e. those that undergo splicing, require in addition the auxiliary function of Los1p. Second, are all 20 aminoacyl-tRNA synthetases present in the nuclear interior and, if so, how are they imported from the cytoplasm? Minor pools of some synthetases have been located in the nuclei of higher eukaryote cell systems (Kisselev and Wolfson 1994), putative NLSs have been identified in several yeast synthetases by computer analysis (Schimmel and Wang 1999), and human MetRS has recently been reported to enter the nucleolus in proliferating cells (Ko et al. 2000). However, a detailed analysis of the intracellular localization of most of the synthetases and their means of crossing the nuclear envelope is still missing. Accordingly, the attractive notion that the synthetases themselves (or a subset thereof) may act as shuttling tRNA export receptors remains speculative at this time. Third, how is the aminoacylation status of a given tRNA recognized by the transport machinery? One has to assume the involvement of a protein that can discriminate between charged and uncharged tRNAs by binding only to the former. If Los1p cannot fulfil this role (see above), one obvious other candidate is the major elongation factor eEF-1A (previously called EF-1α), the eukaryotic homologue of EF-Tu that binds in a GTP-dependent manner to all but one aminoacyl-tRNAs and delivers them to the A-site of elongating ribosomes (Negrutskii and El'skaya 1998). Only initiator methionyl-tRNA^{Met} is not bound by eEF-1 A but rather by the initiation factor eIF-2γ, which takes it to the small ribosomal subunit to form the 43S pre-initiation complex.

7 Nuclear tRNA Export as a Regulatory Mechanism?

Interestingly, both eEF-1A and eIF-2γ have been implicated in nuclear tRNA export through their genetic interaction with Los1p: mutants of *GCD11*, the gene coding for eIF-2γ, are synthetically lethal with the disruption of *LOS1* (Hellmuth et al. 1998), and overexpression of eEF-1A suppresses a synthetically lethal *los1* mutant (Großhans et al. 2000a). Moreover, mutants with reduced levels of, or mutated, eEF-1A (but not eIF-2γ) were found to accumulate tRNA in the nucleus. This defect is synergistic with the *los1*$^-$ mutation, as a *tef2*$^-$ *los1*$^-$ double disruption mutant (an identical protein of eEF-1A is encoded by two genes in the yeast genome, *TEF1* and *TEF2*, either of which is sufficient to sustain cell growth) was shown not only to exhibit a slow-growth phenotype, but also an exacerbated tRNA export defect (Großhans et al. 2000a). However, blocking of translation by other means, e.g., by addition of cycloheximide, did not result in nuclear accumulation of tRNA (Großhans et al. 2000a). Interestingly, communication between the nuclear tRNA export and the translation machinery does not seem to be unilateral. It has recently been found that several conditions that impair nuclear export of tRNA, such as over expression of the tRNA modification enzyme Pus4p, a mutation in $tRNA^{Val}$ or the *los1* disruption, de-repress the translation of *GCN4*, the transcriptional activator of amino acid biosynthetic enzymes. This de-repression is independent of the *GCN2*-encoded kinase or the amount of functional initiator-$tRNA^{Met}$, which normally regulates the level of Gcn4p, and may be caused by a reduction in the efficiency of translation initiation (Qiu et al. 2000). Taken together, these data suggest that the nuclear tRNA export machinery can sense, and react to, defects in protein biosynthesis and vice versa. The amount of mature tRNA inside the nucleus may then act as a signal for coordinating the rate of translation in the cytoplasm with the nuclear events leading to the production and export of tRNA. eEF-1A, which in several cases has been reported to be also found inside the nucleus (Barbarese et al. 1995; Billaut-Mulot et al. 1996; Sanders et al. 1996; Gangwani et al. 1998), may be one of the mediators of this communication between nucleus and cytoplasm. Other factors might include tRNA synthetases, such as the lysyl- and the isoleucyl-tRNA synthetase, which are also involved in the general control network of amino acid biosynthesis (Meussdoerffer and Fink 1983; Lanker et al. 1992). This would also explain the accumulation of non-cognate tRNA upon heat-inactivation of the isoleucyl-tRNA synthetase (see above).

8 Concluding Remarks

Significant progress has been made in recent years towards the understanding of nuclear export of tRNA in both yeast and higher eukaryotes. A transport pathway has been characterized involving a specialized exportin (Xpo-t/Los1p) that can interact directly with tRNA and mediate its translocation across the nuclear pores. The importance of this pathway may, however, vary, depending on the organism. At least in yeast, an additional, Los1p-independent tRNA export pathway appears to be in operation. This is not very surprising as redundancy is often the rule rather than the exception in critical cellular pathways. Although the nature of the alternative pathway is still not clear, it most likely involves aminoacylation of tRNA and other components of the translation machinery. Major translation factors and aminoacyl-tRNA synthetases, which are both ancestral and abundant tRNA-binding proteins, may actually have been the first components to be recruited for nuclear export of tRNA when the processes of transcription and translation became separated spatially by the nuclear envelope. The subsequent evolution of a sophisticated nucleocytoplasmic transport machinery may then have led to the appearance of an exportin specialized for tRNA such as Los1p. In higher eukaryotic organisms, such as *Xenopus laevis*, Xpo-t, the homologue of Los1p, may actually have become the major mediator of nuclear tRNA export, tRNA aminoacylation playing only an auxiliary role. Gene knock-out studies should finally reveal this possibility.

References

Adam SA (1999) Transport pathways of macromolecules between the nucleus and the cytoplasm. Curr Opin Cell Biol 11:402–406

Amberg DC, Goldstein AL, Cole CN (1992) Isolation and characterization of *RAT1*: an essential gene of *Saccharomyces cerevisiae* required for the efficient nucleocytoplasmic trafficking of mRNA. Genes Dev 6:1173–1189

Arts GJ, Fornerod M, Mattaj IW (1998a) Identification of a nuclear export receptor for tRNA. Curr Biol 8:305–314

Arts GJ, Kuersten S, Romby P, Ehresmann B, Mattaj IW (1998b) The role of exportin-t in selective nuclear export of mature tRNAs. EMBO J 17:7430–7441

Azuma Y, Dasso M (2000) The role of Ran in nuclear function. Curr Opin Cell Biol 12:302–307

Barbarese E, Koppel DE, Deutscher MP, Smith CL, Ainger K, Morgan F, Carson JH (1995) Protein translation components are colocalized in granules in oligodendrocytes. J Cell Sci 108: 2781–2790

Becker J, Melchior F, Gerke V, Bischoff FR, Ponstingl H, Wittinghofer A (1995) RNA1 encodes a GTPase-activating protein specific for Gsp1p, the Ran/TC4 homologue of *Saccharomyces cerevisiae*. J Biol Chem 270:11860–11865

Billaut-Mulot O, Fernandez-Gomez R, Loyens M, Ouaissi A (1996) *Trypanosoma cruzi* elongation factor 1-alpha: nuclear localization in parasites undergoing apoptosis. Gene 174:19–26

Boelens WC, Palacios I, Mattaj IW (1995) Nuclear retention of RNA as a mechanism for localization. RNA 1:273–283

Booth JW, Belanger KD, Sannella MI, Davis LI (1999) The yeast nucleoporin Nup2p is involved in nuclear export of importin alpha/Srp1p. J Biol Chem 274:32360–32367

Cheng Y, Dahlberg JE, Lund E (1995) Diverse effects of the guanine nucleotide exchange factor RCC1 on RNA transport. Science 267:1807–1810

Cleary JD, Mangroo D (2000) Nucleotides of the tRNA D-stem that play an important role in nuclear-tRNA export in *Saccharomyces cerevisiae*. Biochem J 347:115–122

Cortes A, Azorin F (2000) DDP1, a heterochromatin-associated multi-KH-domain protein of *Drosophila melanogaster*, interacts specifically with centromeric satellite DNA sequences. Mol Cell Biol 20:3860–3869

Culbertson MR, Winey M (1989) Split tRNA genes and their products: a paradigm for the study of cell function and evolution. Yeast 5:405–427

Dodson RE, Shapiro DJ (1997) Vigilin, a ubiquitous protein with 14 K homology domains, is the estrogen-inducible vitellogenin mRNA 3′-untranslated region-binding protein. J Biol Chem 272:12249–12252

Dworetzky SI, Feldherr CM (1988) Translocation of RNA-coated gold particles through the nuclear pores of oocytes. J Cell Biol 106:575–584

Fahrenkrog B, Hurt EC, Aebi U, Panté N (1998) Molecular architecture of the yeast nuclear pore complex: localization of Nsp1p subcomplexes. J Cell Biol 143:577–588

Gangwani L, Mikrut M, Galcheva-Gargova Z, Davis RJ (1998) Interaction of ZPR1 with translation elongation factor-1alpha in proliferating cells. J Cell Biol 143:1471–1484

Görlich D, Kutay U (1999) Transport between the cell nucleus and the cytoplasm. Annu Rev Cell Dev Biol 15:607–660

Görlich D, Dabrowski M, Bischoff FR, Kutay U, Bork P, Hartmann E, Prehn S, Izaurralde E (1997) A novel class of RanGTP binding proteins. J Cell Biol 138:65–80

Großhans H, Hurt E, Simos G (2000a) An aminoacylation-dependent nuclear tRNA export pathway in yeast. Genes Dev 14:830–840

Großhans H, Simos G, Hurt E (2000b) Transport of tRNA out of the nucleus-direct channeling to the ribosome? J Struct Biol 129:288–294

Hellmuth K, Lau DM, Bischoff FR, Künzler M, Hurt EC, Simos G (1998) Yeast Los1p has properties of an exportin-like nucleocytoplasmic transport factor for tRNA. Mol Cell Biol 18:6374–6386

Her LS, Lund E, Dahlberg JE (1997) Inhibition of Ran guanosine triphosphatase-dependent nuclear transport by the matrix protein of vesicular stomatitis virus. Science 276:1845–1848

Hetzer M, Mattaj IW (2000) An ATP-dependent, Ran-independent mechanism for nuclear import of the U1 A and U2B″ spliceosome proteins. J Cell Biol 148:293–303

Hood JK, Casolari J M, Silver PA (2000) Nup2p is located on the nuclear side of the nuclear pore complex and coordinates Srp1p/importin-α export. J Cell Sci 113:1471–1480

Hopper AK, Martin NC (1992) Processing of yeast cytoplasmic and mitochondrial precursor tRNAs. In: Jones EW, Pringle JR, Broach JR (eds) The molecular and cellular biology of the yeast *Saccharomyces*: gene expression, vol II. Cold Spring Harbor Laboratory, Cold Spring Harbor, pp 99–141

Hopper AK, Banks F, Evangelidis V (1978) A yeast mutant which accumulates precursor tRNAs. Cell 14:211–219

Hopper AK, Schultz LD, Shapiro RA (1980) Processing of intervening sequences: a new yeast mutant which fails to excise intervening sequences from precursor tRNAs. Cell 19:741–751

Hopper AK, Traglia HM, Dunst RW (1990) The yeast RNA1 gene product necessary for RNA processing is located in the cytosol and apparently excluded from the nucleus. J Cell Biol 111:309–321

Hurt DJ, Wang SS, Yu-Huei L, Hopper AK (1987) Cloning and characterization of *LOS1*, a *Saccharomyces cerevisiae* gene that affects tRNA splicing. Mol Cell Biol 7:1208–1216

Izaurralde E, Kutay U, von Kobbe C, Mattaj IW, Görlich D (1997) The asymmetric distribution of the constituents of the Ran system is essential for transport into and out of the nucleus. EMBO J 16:6535–6547

Jarmolowski A, Boelens WC, Izaurralde E, Mattaj IW (1994) Nuclear export of different classes of RNA is mediated by specific factors. J Cell Biol 124:627–635

Kisselev L, Wolfson AD (1994) Aminoacyl-tRNA synthetases from higher eukaryotes. Prog Nucleic Acid Res Mol Biol 48:83–141

Kleeman TA, Wei D, Simpson KL, First EA (1997) Human tyrosyl-tRNA synthetase shares amino acid sequence homology with a putative cytokine. J Biol Chem 272:14420–14425

Ko YG, Kang YS, Kim EK, Park SG, Kim S (2000) Nucleolar localization of human methionyl-tRNA synthetase and its role in ribosomal RNA synthesis. J Cell Biol 149:567–574

Kruse C, Willkomm DK, Grunweller A, Vollbrandt T, Sommer S, Busch S, Pfeiffer T, Brinkmann J, Hartmann RK, Müller PK (2000) Export and transport of tRNA are coupled to a multi-protein complex. Biochem J 346:107–115

Kutay U, Izaurralde E, Bischoff FR, Mattaj IW, Görlich D (1997) Dominant-negative mutants of importin-b block multiple pathways of import and export through the nuclear pore complex. EMBO J 16:1153–1163

Kutay U, Lipowsky G, Izaurralde E, Bischoff FR, Schwarzmaier P, Hartmann E, Görlich D (1998) Identification of a tRNA-specific nuclear export receptor. Mol Cell 1:359–369

Lang BD, Fridovich-Keil JL (2000) Scp160p, a multiple KH-domain protein, is a component of mRNP complexes in yeast. Nucleic Acids Res 28:1576–1584

Lanker S, Bushman JL, Hinnebusch AG, Trachsel H, Mueller PP (1992) Autoregulation of the yeast lysyl-tRNA synthetase gene GCD5/KRS1 by translational and transcriptional control mechanisms. Cell 70:647–57

Lipowsky G, Bischoff FR, Izaurralde E, Kutay U, Schäfer S, Gross HJ, Beier H, Görlich D (1999) Coordination of tRNA nuclear export with processing of tRNA. RNA 5:539–549

Lund E, Dahlberg SG (1998) Proofreading and aminoacylation of tRNAs before export from the nucleus. Science 282:2082–2085

Marck C, Lefebvre O, Carles C, Riva M, Chaussivert N, Ruet A, Sentenac A (1993) The TFIIIB-assembling subunit of yeast transcription factor TFIIIC has both tetratricopeptide repeats and basic helix-loop-helix motifs. Proc Natl Acad Sci USA 90:4027–4031

Melton DA, De Robertis EM, Cortese R (1980) Order and intracellular location of the events involved in the maturation of a spliced tRNA. Nature 284:143–148

Meussdoerffer F, Fink GR (1983) Structure and expression of two aminoacyl-tRNA synthetase genes from *Saccharomyces cerevisiae.* J Biol Chem 258:6293–6299

Motorin Y, Keith G, Simon C, Foiret D, Simos G, Hurt E, Grosjean H (1998) The yeast tRNA:pseudouridine synthase Pus1p displays a multisite substrate specificity. RNA 4:856–869

Nagy E, Rigby WF (1995) Glyceraldehyde-3-phosphate dehydrogenase selectively binds AU-rich RNA in the NAD(+)-binding region (Rossmann fold). J Biol Chem 270:2755–2763

Nakielny S, Dreyfuss G (1999) Transport of proteins and RNAs in and out of the nucleus. Cell 99:677–690

Negrutskii BS, El'skaya AV (1998) Eukaryotic translation elongation factor 1 alpha: structure, expression, functions, and possible role in aminoacyl-tRNA channeling. Prog Nucleic Acid Res Mol Biol 60:47–78

O'Connor JP, Peebles CL (1991) In vivo pre-tRNA processing in *Saccharomyces cerevisiae.* Mol Cell Biol 11:425–439

Panté N, Jarmolowski A, Izaurralde E, Sauder U, Baschong W, Mattaj IW (1997) Visualizing nuclear export of different classes of RNA by electron microscopy. RNA 3:498–513

Pokrywka NJ, Goldfarb DS (1995) Nuclear export pathways of tRNA and 40S ribosomes include both common and specific intermediates. J Biol Chem 270:3619–3624

Popenko VI, Ivanova JL, Cherny NE, Filonenko VV, Beresten SF, Wolfson AD, Kisselev LL (1994) Compartmentalization of certain components of the protein synthesis apparatus in mammalian cells. Eur J Cell Biol 65:60–69

Powers MA, Forbes DJ, Dahlberg JE, Lund E (1997) The vertebrate GLFG nucleoporin, Nup98, is an essential component of multiple RNA export pathways. J Cell Biol 136:241–250

Qiu H, Hu C, Anderson J, Bjork GR, Sarkar S, Hopper AK, Hinnebusch AG (2000) Defects in tRNA processing and nuclear export induce GCN4 translation independently of phosphorylation of the alpha subunit of eukaryotic translation initiation factor 2. Mol Cell Biol 20:2505–2516

Quevillon S, Agou F, Robinson JC, Mirande M (1997) The p43 component of the mammalian multi-synthetase complex is likely to be the precursor of the endothelial monocyte-activating polypeptide II cytokine. J Biol Chem 272:32573–32579

Rout MP, Aitchison JD, Suprapto A, Hjertaas K, Zhao Y, Chait BT (2000) The yeast nuclear pore complex: composition, architecture, and transport mechanism. J Cell Biol 148:635–651

Ryan KJ, Wente SR (2000) The nuclear pore complex: a protein machine bridging the nucleus and cytoplasm. Curr Opin Cell Biol 12:361–371

Sanders J, Brandsma M, Janssen GM, Dijk J, Moller W (1996) Immunofluorescence studies of human fibroblasts demonstrate the presence of the complex of elongation factor-1 beta gamma delta in the endoplasmic reticulum. J Cell Sci 109:1113–1117

Sarkar S, Hopper AK (1998) tRNA nuclear export in *Saccharomyces cerevisiae*: In situ hybridization analysis. Mol Biol Cell 9:3041–3055

Sarkar S, Azad AK, Hopper AK (1999) Nuclear tRNA aminoacylation and its role in nuclear export of endogenous tRNAs in *Saccharomyces cerevisiae.* Proc Natl Acad Sci USA 96:14366–14371

Schimmel P, Wang CC (1999) Getting tRNA synthetases into the nucleus. Trends Biochem Sci 24:127–128

Sharma K, Fabre E, Tekotte H, Hurt EC, Tollervey D (1996) Yeast nucleoporin mutants are defective in pre-tRNA splicing. Mol Cell Biol 16:294–301

Shen WC, Selvakumar D, Stanford DR, Hopper AK (1993) The *Saccharomyces cerevisiae LOS1* gene involved in pre-tRNA splicing encodes a nuclear protein that behaves as a component of the nuclear matrix. J Biol Chem 268:19436–19444

Shen WC, Stanford DR, Hopper AK (1996) Los1p, involved in yeast pre-tRNA splicing, positively regulates members of the *SOL* gene family. Genetics 143:699–712

Simos G (1999) Nuclear export of tRNA. Protoplasma 209:173–180

Simos G, Segref A, Fasiolo F, Hellmuth K, Shevchenko A, Mann M, Hurt EC (1996a) The yeast protein Arc1p binds to tRNA and functions as a cofactor for methionyl- and glutamyl-tRNA synthetases. EMBO J 15:5437–5448

Simos G, Tekotte H, Grosjean H, Segref A, Sharma K, Tollervey D, Hurt EC (1996b) Nuclear pore proteins are involved in the biogenesis of functional tRNA. EMBO J 15:2270–2284

Simos G, Sauer A, Fasiolo F, Hurt EC (1998) A conserved domain within Arc1p delivers tRNA to aminoacyl-tRNA synthetases. Mol Cell 1:235–242

Singh R, Green MR (1993) Sequence-specific binding of transfer RNA by glyceraldehyde-3-phosphate dehydrogenase. Science 259:365–368

Sträßer K, Hurt E (1999) Nuclear RNA export in yeast. FEBS Lett 452:77–81

Talcott B, Moore MS (1999) Getting across the nuclear pore complex. Trends Cell Biol 9:312–318

Tobian JA, Drinkard L, Zasloff M (1985) tRNA nuclear transport: defining the critical region of human $tRNA^{Met}$ by point mutagenesis. Cell 43:415–422

Weber V, Wernitznig A, Hager G, Harata M, Frank P, Wintersberger U (1997) Purification and nucleic-acid-binding properties of a *Saccharomyces cerevisiae* protein involved in the control of ploidy. Eur J Biochem 249:309–317

Wente SR (2000) Gatekeepers of the nucleus. Science 288:1374–1377

Westaway SK, Abelson J (1995) Splicing of tRNA Precursors. In: Söll D, RajBhandary UL (eds) tRNA: Structure, biosynthesis, and function. ASM Press Washington, DC, pp 79–92

Wilhelm ML, Reinbolt J, Gangloff J, Dirheimer G, Wilhelm FX (1994) Transfer RNA binding protein in the nucleus of *Saccharomyces cerevisiae.* FEBS Lett 349:260–264

Wilson R, Ainscough R, Anderson K, Baynes C, Berks M, Bonfield J, Burton J, Connell M, Copsey T, Cooper J, et al. (1994) 2.2 Mb of contiguous nucleotide sequence from chromosome III of *C. elegans.* Nature 368:32–38

Wolin SL, Matera AG (1999) The trials and travels of tRNA. Genes Dev 13:1–10

Yan W, Schilke B, Pfund C, Walter W, Kim S, Craig EA (1998) Zuotin, a ribosome-associated DnaJ molecular chaperone. EMBO J 17:4809–4817

Zasloff M (1983) tRNA transport from the nucleus in eukaryotic cell: carrier-mediated process. Proc Natl Acad Sci USA 80:6436–6440

Zhang S, Lockshin C, Herbert A, Winter E, Rich A (1992) Zuotin, a putative Z-DNA binding protein in *Saccharomyces cerevisiae.* EMBO J 11:3787–3796

Nuclear Export of Messenger RNA

Elisa Izaurralde[1]

1 Introduction

Messenger RNAs (mRNAs) exist in the cell in dynamic association with multiple distinct proteins. The mRNA export cargo is therefore a ribonucleoprotein particle (mRNP). The most striking example of this are the Balbiani ring (BR) mRNPs of *Chironomus tentans*, which are composed of an RNA molecule of 35–40 kb in length associated with about 500 protein molecules (Mehlin et al. 1992; reviewed by Daneholt 1997). Among the most abundant RNA-binding proteins found associated to heterogeneous (in size) nuclear mRNPs are the hnRNP proteins. There are about 20 different hnRNP proteins in higher eukaryotes (hnRNP A to hnRNP U, reviewed by Dreyfuss et al. 1993).

A general feature of mRNP export is that it is an active, receptor-mediated process. Since the major classes of exported cellular RNAs (i.e. mRNAs, tRNA, ribosomal RNAs and the polymerase II-transcribed spliceosomal U snRNAs) do not compete with each other for export, mRNPs are thought to utilize distinct and specific saturable export mediators not shared by other classes of cellular RNAs (Bataillé et al. 1990; Jarmolowski et al. 1994; Pokrywka and Goldfarb 1995).

mRNP export occurs through nuclear pore complexes (NPCs). NPCs are large proteinaceous structures that form aqueous channels across the nuclear envelope (reviewed by Fahrenkrog et al., this Vol.). NPCs are composed of multiple copies of up to about 30–60 distinct polypeptides called nucleoporins. Nucleoporins are characterized by the presence of protein/protein interaction domains and they also typically contain clusters of multiple phenylalanine-glycine dipeptide repeats. It has been proposed that these FG-repeats provide docking sites during translocation of receptor-cargo complexes through the central channel of the pore (Fahrenkrog et al., this Vol.). Therefore, mRNP export receptors are expected to interact with multiple nucleoporins.

Nuclear export of cellular mRNPs is highly selective, because only fully processed RNAs are normally exported. Incompletely spliced precursor mRNAs (pre-mRNAs) and excised introns are actively retained within the nucleus. Studies in yeast and in vertebrate cells suggested that pre-mRNAs are

[1] EMBL, Meyerhofstrasse 1, 69117 Heidelberg, Germany

Results and Problems in Cell Differentiation, Vol. 35
K. Weis (Ed.): Nuclear Transport

held within the nucleus by association with spliceosomal factors (Legrain and Rosbash 1989; Hamm and Mattaj 1990; Rutz and Séraphin 2000). Electron microscopy studies on the export of Balbiani ring mRNPs have shown that several proteins associated with the mRNP in the nucleoplasm are selectively removed before or during translocation across the NPC (reviewed by Daneholt 1997). Together, these observations imply that the mRNA export pathway may include steps at which specific retention factors are removed, by an as yet unknown mechanism.

Proteins implicated in the export of cellular mRNA include several nucleoporins (Fahrenkrog et al., this Vol.) and RNA-binding proteins, the RNA helicase Dbp5, and the NPC-associated proteins hRAE1/Gle2p and TAP/Mex67p. Although a contribution of the importin-β-related nuclear transport receptors to mRNP export cannot be ruled out (Fornerod and Ohno, this Vol.), several lines of evidence indicate that hRAE1/Gle2p and TAP/Mex67p may fulfill the function of export receptors for cellular mRNPs. This review focuses on the role of these putative export receptors, of Dbp5 and of several RNA-binding proteins in the export of mRNA.

2 RAE1/Gle2p and TAP/Mex67p May Function as Nuclear Export Receptors for mRNA

The yeast proteins Gle2p (also known as Rae1p) and Mex67p were originally identified in searches for mutants that accumulate polyadenylated RNAs within the nucleus or in screens for genetic interactions with nucleoporins. Although these proteins do not exhibit sequence similarities, they have both been implicated in mRNA export. Mex67p is essential for mRNA export in *Saccharomyces cerevisiae*, and Gle2p is essential in *Schizosaccharomyces pombe* (Brown et al. 1995; Murphy et al. 1996; Segref et al. 1997). The vertebrate homologues of Mex67p and Gle2p, TAP (also known as NXF1) and RAE1 (also known as mrnp41) respectively, have also been implicated in the export of cellular mRNA (see below). In this review, human NXF1 will be called TAP, human RAE1 will be referred to as hRAE1 and its yeast homologue Gle2p.

hRAE1/Gle2p and TAP/Mex67p have several of the characteristic features of nuclear export receptors. Nuclear export receptors are expected to interact with components of the NPC, to shuttle between the nucleus and cytoplasm and to bind directly or indirectly to their cargoes. In vertebrate and yeast cells hRAE1/Gle2p and TAP/Mex67p have been visualized at both the nucleoplasmic and cytoplasmic faces of the NPC (Kraemer and Blobel 1997; Santos-Rosa et al. 1998; Bachi et al. 2000). This is consistent with in vitro and genetic interactions of these proteins with multiple nucleoporins (Brown et al. 1995; Murphy et al. 1996; Segref et al. 1997; Bailer et al. 1998; Santos-Rosa et al. 1998; Katahira et al. 1999; Pritchard et al. 1999; Bachi et al. 2000; Strawn et al. 2000). In addition, TAP and hRAE1 shuttle between the nuclear and cytoplasmic compartments (Bear et al. 1999; Braun et al. 1999; Kang and Cullen 1999; Katahira et al. 1999; Pritchard et al. 1999). Finally, these proteins must bind mRNPs directly or indirectly because they can be cross-linked to polyadenylated RNAs (Kraemer and Blobel 1997; Santos-Rosa et al. 1998; Katahira et al. 1999). Thus, TAP/Mex67p and hRAE1/Gle2p may mediate mRNP/NPC interactions during export. Consequently, these proteins may perform the function of mRNP export receptors distinct from the importin-β-related family of Ran binding proteins (reviewed by Jäckel and Görlich, this Vol.).

2.1 Role of hRAE1/Gle2p in mRNP Nuclear Export

As mentioned above, the yeast and human RAE1/Gle2p proteins have been shown to interact with nucleoporins. Gle2p binds Nup116p directly in *S. cerevisiae* (Bailer et al. 1998), while in *S. pombe* a functional interaction with Npp106p, a nucleoporin similar to *S. cerevisiae* Nic96p, has been described (Yoon et al. 1997). Human RAE1 interacts directly with the nucleoporin Nup98 (Pritchard et al. 1999). However, apart from its binding to Nup98, hRAE1 is likely to directly or indirectly interact with other components of the NPC for the following reasons. First, Nup98 localizes on the nucleoplasmic side of the pore while hRAE1 has been visualized on both faces of the NPC (Kraemer and Blobel 1997). Second, in *Nup98-/-* cells, in which Nup98 is no longer expressed, hRAE1 is still detected at the nuclear rim (Kasper et al. 2001). The association of human RAE1 with the NPC requires ongoing transcription, indicating that the protein establishes a functional, but not a permanent association with the NPC (Prichard et al. 1999). This conclusion received strong support from studies on the export of BR mRNPs in *C. tentans* (Sabri and Visa 2000). In this system, RAE1 did not bind to the BR particles either cotranscriptionally or in the nucleoplasm but at the NPC, just before translocation. After translocation, RAE1 was no longer bound to the cytoplasmic particles, suggesting that the protein remains at the NPC or is rapidly dissociated from the exported mRNP at the cytoplasmic side of the pore. Moreover, the localization of RAE1 at the NPC correlated with the presence of an exported mRNP in the pore.

RAE1 not only associates with BR mRNPs but can be cross-linked to polyadenylated RNAs (Kraemer and Blobel 1997), indicating that the protein binds mRNA directly or via additional proteins. Intriguingly, in vitro, hRAE1 binds to TAP and an interaction between the *S. pombe* proteins Gle2p and Mex67p has been reported (Yoon et al. 2000). Although it is not yet clear whether these interactions are direct, these data suggest that the two proteins may cooperate to export cellular mRNAs (Bachi et al. 2000; Yoon et al. 2000). Thus, understanding the function of hRAE1/Gle2 in mRNA export awaits further studies and, more importantly, the identification of its cellular partners.

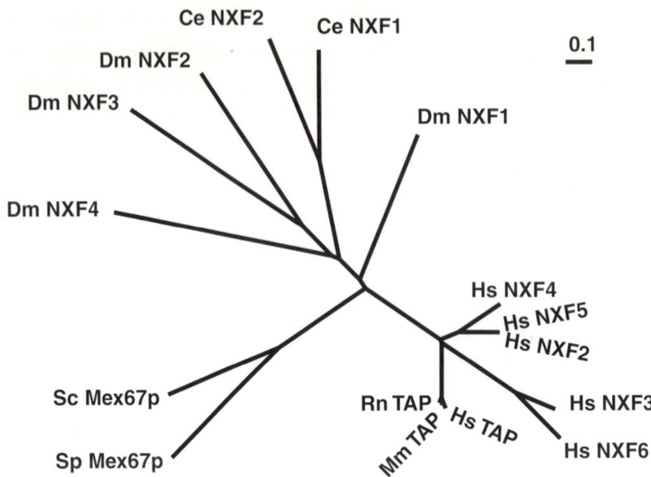

Fig. 1. Phylogenetic tree of NXF family sequences. Ce, *Caenorhabditis elegans*; Dm, *Drosophila melanogaster*; Hs, *Homo sapiens*; Mm, *Mus musculus*; Rn, *Rattus norvegicus*; Sc, *Saccharomyces cerevisiae*; Sp, *Schizosaccharomyces pombe*

2.2 TAP and Mex67p Belong to the Conserved Family of Nuclear Export Factor Proteins

Mex67p and human TAP belong to an evolutionarily conserved family of proteins, referred to as the nuclear export factor (NXF) family, which has two or more members in higher eukaryotes (Fig. 1; Herold et al. 2000). There are two NXF proteins in *Caenorhabditis elegans*, four in *Drosophila melanogaster* and six putative homologues in human. The phylogenetic tree shown in Fig. 1 indicates that the multiplication of *nxf* genes has occurred independently in different lineages. This diversification of NXF proteins in higher eukaryotes as compared to yeast may reflect an adaptation to an increasing number and complexity of mRNA export cargoes or, more probably, to tissue specific requirements. Consistent with this latter possibility is the observation that human NXF5 is mainly expressed in brain (Lin et al., pers. comm.).

Evidence for a role in mRNP export has recently been obtained for *C. elegans* and *D. melanogaster* NXF1 (Tan et al. 2000; A. Herold and E. Izaurralde, unpubl. data) and for human TAP (human NXF1; Braun et al. 2001; Guzik et al. 2001). Moreover, TAP has been implicated in the export of simian type D retroviral RNAs bearing the constitutive transport element (CTE; Grüter et al. 1998). Based on the structural and functional conservation of the NXF proteins and the observation that human TAP can partially substitute Mex67p (Katahira et al. 1999), it is likely that most, if not all, members of the NXF family play a role in mRNA export.

Members of the NXF family are characterized by the presence of several conserved structural domains. From the N to the C-terminus these are: an

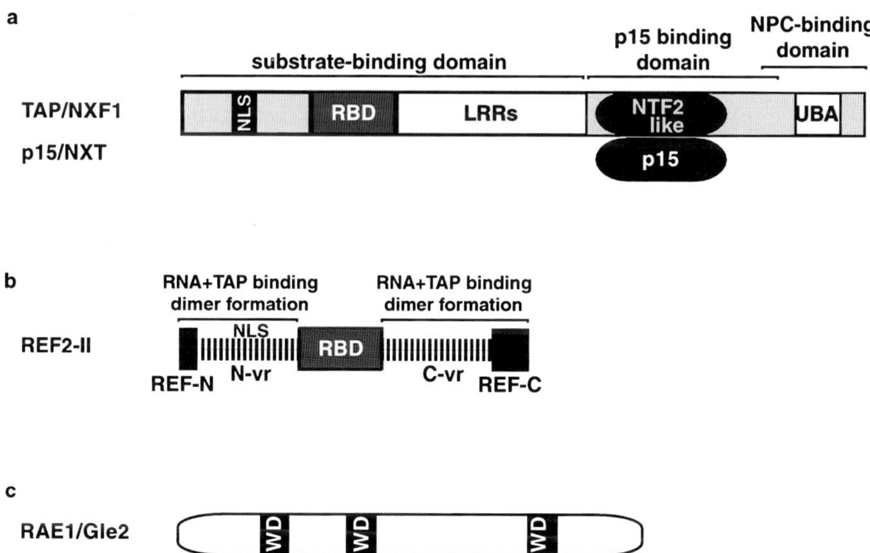

Fig. 2a–c. Domain organization of proteins implicated in mRNA nuclear export. **a** TAP/p15 heterodimers. These proteins are also known as NXF1/NXT. *NLS* Nuclear localization signal, *RBD* RNA-binding domain found only in human homologues and *D. melanogaster* and *C. elegans* NXF1, *LRRs* leucine-rich repeat domain, *NTF2-like* NTF2-like domain, *UBA* ubiquitin-associated domain. The functional role of the domains is as described by Bachi et al. (2000) and Herold et al. (2000). p15/NXT is related to NTF2 and heterodimerizes with the NTF2-like domain of NXFs. **b** REF proteins. Domain organization of Yra1/REF proteins as described by Stutz et al. (2000) and Rodrigues et al. (2001). *RBD* RNA-binding domain, *REF-N* and *REF-C* conserved N- and C-terminal motifs. *N-vr* and *C-vr* represent the N-and C-terminal variable regions specific to each member of the family. The regions mediating RNA, TAP binding and dimerization of these proteins are indicated. **c** The RAE1/Gle2p contains three WD-domains. (Brown et al. 1995; Murphy et al. 1996; Kraemer and Blobel 1997)

RNA-binding domain (RBD), a leucine-rich repeat (LRR) domain, a middle region showing significant sequence similarity to the nuclear transport factor 2 (the NTF2-like domain) and a C-terminal region related to ubiquitin-associated (UBA) domains (the UBA-like domain; Fig. 2; Herold et al. 2000; Liker et al. 2000; Suyama et al. 2000). The LRR and the NTF2-like domains are the most conserved features of NXF proteins, whereas the RBD and the UBA-like domain are not always present. The function of these domains is discussed below.

2.2.1 NXF Proteins Act as Heterodimers

The NTF2-like domain is conserved in most members of the NXF-family. In higher eukaryotes this domain allows heterodimerization with a small nuclear protein know as p15/NXT (Suyama et al. 2000). p15 is also related to NTF2

(Black et al. 1999; Katahira et al. 1999) but while NTF2 forms homodimers, p15 heterodimerizes with the NTF2-like domain of NXF proteins. Since there is only one *p15* gene in *D. melanogaster* and *C. elegans* and two in available human genomic sequences (p15–1 and p15–2, Herold et al. 2000), this suggests that one p15 protein may heterodimerize with various NXF proteins. Indeed, both human p15–1 and p15–2 interact with TAP and human NXF2, NXF3 and NXF5 (Herold et al. 2000; Lin et al., pers. comm.).

The role of p15 in mRNP export is not well-defined. p15 binding to TAP is not strictly required for TAP-mediated export of CTE-bearing substrates (Bachi et al. 2000; Kang et al. 2000) but is required for the mRNA export activity of the NXF proteins (Herold et al. 2000; Braun et al. 2001; Guzik et al. 2001). p15 has been reported to interact with RanGTP and to participate in multiple export pathways including CRM1-mediated export (Black et al. 1999, 2001), but these observations could not be reproduced in other laboratories (Katahira et al. 1999; Herold et al. 2000; Braun et al. 2001; Guzick et al. 2001). Our data suggest that p15 participates in mRNA export by heterodimerizing with TAP, or other members of the NXF family, and has no intrinsic export activity (Braun et al. 2001). Further studies are undoubtedly required to understand p15 function.

The NTF2-like domain also occurs in *S. pombe* and *S. cerevisiae* Mex67p (Suyama et al. 2000). However, the functional role of this domain in Mex67p is unclear as there is no obvious p15 homologue encoded by the yeast genome. In *S. cerevisiae* Mex67p interacts with a protein known as Mtr2p (Santos Rosa et al. 1998). Although Mtr2p does not exhibit obvious sequence similarities with p15, secondary structure predictions suggest that it could be a p15 analogue and may therefore interact with the NTF2-like domain of Mex67p (Sträßer et al. 2000; Suyama et al. 2000). Furthermore, the observation that co-expression of human TAP and p15–1 in yeast partially restores growth of a strain carrying the otherwise lethal *mex67/mtr2* double-knockout suggests that Mtr2p and p15 may have a similar function (Katahira et al. 1999).

2.2.2 Interaction of NXF Proteins with the NPC

TAP interacts in vitro with multiple FG-containing nucleoporins including CAN, Nup98, p62, Nup153 and hCG1 (Katahira et al. 1999; Bachi et al. 2000). Binding of TAP to nucleoporins is mediated by its C-terminal NPC-binding domain (Fig. 2; Bachi et al. 2000). In HeLa cells the NPC-binding domain of TAP was shown to be necessary and sufficient for the localization of the protein to the nuclear envelope (Bear et al. 1999; Bachi et al. 2000). Furthermore, this domain is a potent inhibitor of multiple export pathways including export mediated by CRM1 and exportin-t (Bachi et al. 2000). A similar dominant negative effect of the NPC-binding domain of importin-β on multiple transport pathways has been described (Kutay et al. 1997) and indicates that, in vivo, the NPC-binding domains of transport receptors compete with each other for binding sites at the NPC. Thus, the interaction of TAP with multiple FG-repeat-containing nucleoporins is analogous to interactions described between nucleoporins and the transport receptors of the importin-β-like family. However, while the interactions between nucleoporins and the importin-β-like receptors are regulated by Ran (see chapters by Jäckel and Görlich, and Fornerod and Ohno, this Vol.), the mechanism by which binding of TAP to the NPC is regulated is unknown.

The nucleoporin-binding domain of TAP contains the UBA-like domain (Fig. 2; Suyama et al. 2000). The structure of UBA domains (Dieckmann et al. 1998) suggests that residues located in a conserved loop play a critical role in these interactions. Indeed, substitution of these residues by alanines blocks the ability of TAP to promote export of both cellular mRNA (Braun et al. 2001) and CTE-bearing RNAs (Kang and Cullen 1999; Herold et al. 2000). These mutations also prevent binding of TAP and human NXF2 to nucleoporins in vitro and in vivo (Herold et al. 2000). The UBA domain is conserved in yeast Mex67p but only the *S. cerevisiae* protein has been shown to interact directly with nucleoporins (Strawn et al. 2000). As is the case for the metazoan proteins, Mex67p lacking the UBA domain no longer localizes to the nuclear envelope (Sträßer et al. 2000). Previously, it had been reported that binding to Mtr2p was required for the association of Mex67p with the NPC, because a Mex67p mutant that no longer interacts with Mtr2p did not localize to the nuclear envelope (Santos Rosa et al. 1998). However, this mutation could also have directly affected nucleoporin binding or Mex67p folding. The more recent observation, that deleting the UBA-like domain of Mex67p without affecting Mtr2p binding prevents its association with the NPC in vivo, suggests that the mode of interaction of yeast and metazoan NXF proteins with nucleoporins is more conserved than previously thought.

2.2.3 Binding of NXF Proteins to mRNPs

Binding of TAP to viral mRNPs bearing the CTE is direct and mediated by its N-terminal domain (Grüter et al. 1998; Braun et al. 1999; Kang and Cullen 1999; Liker et al. 2000). The mode of interaction with cellular mRNPs remains to be established, but is probably different from that with the CTE RNA for the following reasons. First, most cellular mRNAs do not contain sequences similar to those present in the CTE. Second, although TAP exhibits general affinity for RNAs (Grüter et al. 1998; Braun et al. 1999; Katahira et al. 1999; Stutz et al. 2000), its affinity for the CTE RNA is approximately three orders of magnitude higher than that for the DHFR mRNA in vitro (Braun et al. 1999). Third, CTE-bearing RNAs coimmunoprecipitate with TAP in the absence of cross-links (Braun et al. 1999), whereas copurification of TAP and Mex67p with polyadenylated RNAs is only observed after extensive UV-irradiation (Santos Rosa et al. 1998; Katahira et al. 1999). Fourth, while an excess of CTE RNA saturates export of mRNAs in *Xenopus laevis* oocytes, an excess of mRNA does not interfere with

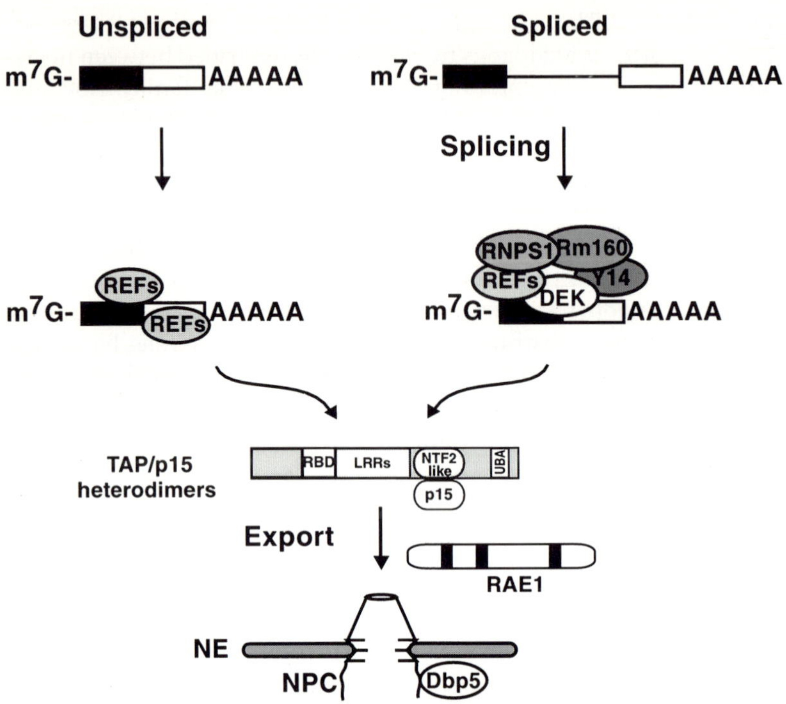

Fig. 3. The mRNA export pathway. Proteins implicated in mRNA export directly include the hnRNP-like REF proteins (Yra1p in yeast and Aly in mouse), TAP/p15 heterodimers (or Mex67p/Mtr2p in yeast) and RAE1 (also known as Gle2p). These proteins interact with spliced or unspliced mRNAs but also between themselves most likely to form the mRNP fiber. The precise order of association of these proteins with the mRNA is unknown. SRm160, DEK, Y14, REFs and RNPS1 preferentially associate with spliced mRNAs, and may facilitate the recruitment of TAP or RAE1 after splicing. Note that direct evidence for a role of SRm160, DEK, Y14 and RNPS1 in mRNP export is lacking. In the cytoplasmic side of the NPC, the DEAD box helicase Dbp5 may play an essential role in dissociating proteins as the mRNP emerges from the pore. RNA-binding proteins may be recycled back to the nucleus or may remain associated with the mRNP in the cytoplasm

CTE export (Pasquinelli et al. 1997; Saavedra et al. 1997). Based on these observations, we have proposed that the CTE bypasses several steps in the mRNA export pathway by directly interacting with TAP, while cellular mRNAs may recruit TAP via protein-protein interactions (Fig. 3; Grüter et al. 1998; Braun et al. 1999; Bachi et al. 2000). Recently, several TAP/Mex67p partners that could facilitate their interaction with cellular mRNPs have been identified. These include E1B-AP5 and members of an evolutionarily conserved family of hnRNP-like proteins, the Yra1p/REF proteins (Bachi et al. 2000; Sträßer and Hurt 2000; Stutz et al. 2000).

The interaction of TAP with E1B-AP5 and members of the REF family of proteins is mediated by its N-terminal domain (Bachi et al. 2000; Stutz et al.

2000). This domain is also responsible for general RNA-binding affinity and specific binding to the CTE RNA (Braun et al. 1999; Kang and Cullen 1999; Liker et al. 2000) suggesting that it functions as a cargo-binding domain. This domain comprises the LRRs, the RBD and the more divergent sequences upstream of the RBD (Fig. 2; Herold et al. 2000). These poorly conserved sequences may confer different substrate specificities to the different family members. Consistently, human NXF2 and NXF3 both interact with E1B-AP5, but only NXF2 binds REF proteins (Herold et al. 2000).

3 Role of hnRNP-Like Proteins in mRNA Export

In yeast, several proteins structurally related to vertebrate hnRNP proteins have been identified (Wilson et al. 1994). For instance, Npl3p/Nop3p is structurally and functionally related to vertebrate hnRNP A1 (Flach et al. 1994; Wilson et al. 1994), and mutations in Npl3 that affect the export of polyadenylated mRNA from the nucleus have been described (Russel and Tollervey 1995; Singleton et al. 1995; Lee et al. 1996). Like many of the major vertebrate hnRNP proteins, Npl3 shuttles between nucleus and cytoplasm and accompanies the mRNP during export. However the question of whether Npl3 or vertebrate hnRNP proteins directly contribute to the export process remains unanswered.

In addition to the major hnRNP proteins (A1–U), several RNA-binding proteins playing a role in mRNP export have been described recently. These proteins share with hnRNP proteins many of the structural domains (RBDs, RGG-boxes, etc.). However, in contrast to hnRNPs A1–U, they are not present in complexes isolated by coimmunuprecipitation with anti-hnRNP C antibodies and are therefore generically referred to as hnRNP-like proteins.

3.1 hnRNP-Like Proteins May Recruit NXFs to mRNP Complexes

The hnRNP-like proteins E1B-AP5 and REFs interact with TAP in vitro and may facilitate its recruitment to the mRNPs (Bachi et al. 2000; Stutz et al. 2000). E1B-AP5 was originally identified in a screen for cellular proteins interacting with the adenovirus protein E1B-55, and shows extensive homology with hnRNP U and other hnRNP proteins (Gabler et al. 1998). The detailed mechanism by which this protein participates in mRNP export is unknown. The REF proteins are hnRNP-like proteins characterized by the presence of one RBD and two conserved motifs at their N- and C-termini, the REF-N and REF-C boxes (Fig. 2; Stutz et al. 2000). REF proteins are evolutionarily conserved from yeast to human and are often encoded by more than one related gene per species (Stutz et al. 2000). Yra1p, a *S. cerevisiae* member of the family, is an essential RNA-binding protein (Portman et al. 1997) that interacts directly with TAP and Mex67p and is involved in mRNA nuclear export (Sträßer and Hurt 2000; Stutz et al. 2000). The murine members of the family preferentially interact with TAP (Stutz et al. 2000), but at least one variant (called Aly) can partially restore growth of a Yra1p-deleted strain indicating a functional conservation (Sträßer and Hurt 2000). Furthermore, direct evidence for a role of vertebrate REFs in mRNA export has recently been reported (Rodrigues et al. 2001; Zhou et al. 2000).

Apart from E1B-AP5 and REFs, other RNA binding proteins may also facilitate TAP binding to the mRNPs. This possibility is consistent with the observation that Y14, an hnRNP-like protein that preferentially associates with spliced mRNAs, has recently been shown to be present in a complex containing TAP (Kataoka et al. 2000). Although it is not currently clear whether Y14 interacts directly with TAP, it is possible that Y14 and/or additional proteins present in this complex might recruit TAP to mRNPs in a splicing-dependent manner (see below).

4 Export of Spliced and Unspliced mRNAs

Numerous examples in the literature indicate that expression of specific genes in mammalian cells is strongly stimulated by the presence of an intron, while mRNAs transcribed from the corresponding cDNAs are expressed poorly (Luo et al. 1999 and references therein). This gave rise to the concept that splicing might deposit proteins that influence the transport of spliced mRNAs to the cytoplasm.

Several proteins that associate with mRNAs during splicing and distinguish spliced from unspliced mRNAs have now been identified. These include Y14 (Kataoka et al. 2000; Le Hir et al. 2000a), REF/Aly (Le Hir et al. 2000a; Zhou et al. 2000), RNPS1 (Le Hir et al. 2000a), SRm160 (Le Hir et al. 2000a,b; McCracken et al., pers. comm.) and DEK (Le Hir et al. 2000a; McGarvey et al. 2000). A common feature of these proteins is that they localize within the nucleoplasm, with sites of highest concentration in structures known as speckled domains (Fig. 4). Nuclear speckles are sites of enrichment of splicing factors (reviewed

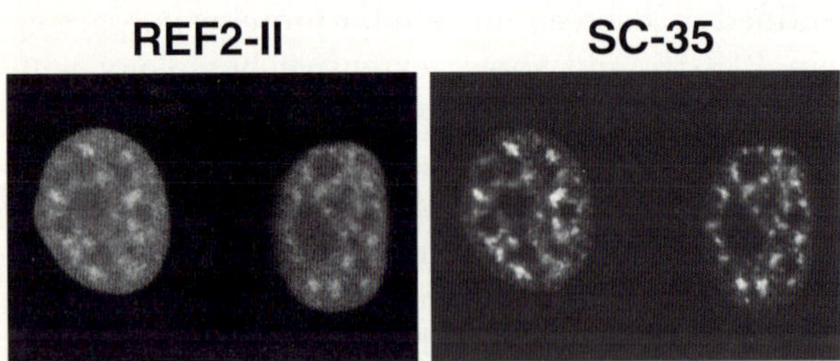

Fig. 4. Subcellular localization of REF2-II in Hela cells. REF2-II is localized within the nucleoplasm, concentrated in speckles. These speckles co-localize precisely with the structures labeled by the antibodies directed to the splicing factor SC-35 (Rodrigues et al. 2001). Similar localization has been reported for Y14 and DEK. (Kataoka et al. 2000; McGarvey et al. 2000)

by Misteli and Spector 1998; Lindsay et al. 2000), so the localization of REF, Y14, SRm160, RNPS1 and DEK in these structures suggests that they are able to interact with components of the splicing machinery, though only SRm160 and RNPS1 have been shown to play a role in splicing (Le Hir et al. 2000a and references therein). Furthermore, these proteins are components of a 340-kDa complex deposited by the spliceosome 20–24 nucleotides upstream of a splice junction in a splicing-dependent, but sequence-independent, manner (Fig. 3, Le Hir et al. 2000a,b). It is not yet clear how such a large complex can bind with high affinity in a sequence-independent manner, but the existence of this complex containing TAP-binding proteins implies that splicing ensures the recruitment of TAP, and hence, the export of spliced mRNAs, in a sequence-independent manner. In contrast, export of unspliced mRNAs would depend on their ability to recruit TAP partners or TAP efficiently, which may in turn depend on their primary sequence and/or length (Braun et al. 2001; Rodrigues et al. 2001). This provides a rationale for the observation that splicing enhances export (and/or expression) of some specific genes in mammalian cells. Indeed, the presence of an intron confers a kinetic advantage to mRNAs generated by splicing, compared to the corresponding cDNA-derived mRNAs, when competing with other mRNAs for export factors. There are no obvious yeast homologues of SRm160, DEK and RNPS1, suggesting that some but not all steps of the mRNA export pathway are conserved.

5 Disassembly of Export Complexes and Recycling of Export Factors: The Role of Dbp5

During mRNP nuclear export, dynamic rearrangements of RNA secondary structure and of RNA/protein interactions are likely to occur. Studies on the translocation of BR RNPs through the NPC illustrate this point (reviewed by Daneholt 1997). During transcription, the BR mRNA associates with multiple proteins and folds into a crescent-shaped structure of about 50 nm in diameter. This particle is partially unfolded into an RNP fiber of 25 nm in diameter before NPC translocation. Studies of the protein composition of BR mRNPs showed that individual proteins are selectively removed at different stages during their formation and passage through the nucleoplasm and NPC into the cytoplasm. Some splicing factors dissociate just after splicing has occurred (Kiseleva et al. 1994). However, this is not the case for all splicing factors as the homologue of the splicing factor ASF/SF2 remains associated with the mRNP in the nucleoplasm and dissociates just before NPC translocation (Alzahnova-Ericsson et al. 1996). Like all polymerase-II-transcribed RNAs, mRNAs have a 5′ cap structure. This structure is recognized by a heterodimeric nuclear cap-binding protein complex (CBC) consisting of a cap-binding protein of 80 kDa (CBP80) and a cap-binding protein of 20 kDa (CBP20) (Izaurralde et al. 1994, 1995). The CBC binds to the 5′ end of the BR mRNP and remains bound to the cap structure during translocation of the RNP, but is removed immediately after translocation (Visa et al. 1996a). This is particularly interesting since BR RNP export occurs with the 5′ end of the RNA interacting with the pore first and leading the way as the RNA translocates into the cytoplasm. Finally, some hnRNP proteins remain associated with the RNA after translocation and are detected on polyribosomes (Visa et al. 1996b). The mechanism by which mRNA-binding proteins associated with the export substrate are specifically removed from the RNA before, during or after NPC translocation is unknown.

A protein that may play an important role in disrupting mRNA/protein interactions as the mRNP emerges from the NPC is Dbp5 (Snay-Hodge et al. 1998; Tseng et al. 1998; Schmitt et al. 1999). Dbp5 belongs to the DEAD-box family of RNA helicases. RNA helicases are enzymes that use energy derived from nucleoside triphosphate hydrolysis to unwind short duplex regions in RNA molecules and may also be implicated in disrupting RNA/protein interactions (reviewed by de la Cruz et al. 1999). In most cases, the NTPase activity is stimulated or is dependent on RNA binding, and therefore RNA helicases are considered as RNA-dependent NTPases. Dbp5 has been directly implicated in export of mRNAs from the nucleus in yeast and vertebrate cells (Snay-Hodge et al. 1998; Tseng et al. 1998; Schmitt et al. 1999). Both in yeast and human cells Dbp5 is mainly cytoplasmic but a fraction of the protein is recruited to the cytoplasmic fibrils of the NPC via a direct interaction with the nucleoporin CAN/Nup159p (Hodge et al. 1999; Schmitt et al. 1999; Strahm et al. 1999). Dbp5 exhibits RNA-dependent ATPase activity, so movement and/or conformational rearrangements of large RNPs during the translocation through the NPC may depend on the hydrolysis of ATP by Dbp5. Independent of the exact mechanism of action of Dbp5, the direct implication of this RNA-dependent ATPase in export may, at least in part, account for the energy requirements and the directionality of this process.

5.1 The Terminal Step of the mRNP Export Process May Occur at the Cytoplasmic Fibrils of the NPC

Nup159p is an essential yeast nucleoporin which, like its vertebrate homologue CAN (Nup214), has been localized to the cytoplasmic fibrils of the NPC and implicated in mRNA export (Kraemer et al. 1994; 1995; Gorsch et al. 1995; Del Priore et al. 1996, 1997; van Deursen et al. 1996, Belgareh et al. 1998; Boer et al. 1998;;; Hurwitz et al. 1998;). The N-terminal domain of Nup159p is not essential for growth, but appears to contribute substantially to its mRNA export function (Del Priore et al. 1997). Indeed, in cells expressing a mutant protein lacking this domain, nuclear accumulation of polyadenylated RNAs was observed even at the permissive temperature, and this defect became more severe following a shift to 37°C. These findings indicate that the N-terminal domain of Nup159p is required for efficient mRNA export (Del Priore et al. 1997). The N-terminal domain of both Nup159p and CAN binds directly to Dbp5 (Schmitt et al. 1999) suggesting that recruitment of Dbp5 to the cytoplasmic fibrils of the NPC increases the efficiency of the export process. Since the N-terminal domain of Nup159p is not essential, in its absence yeast Dbp5p can still accomplish its essential function, probably because of its high concentration in the cytoplasm or its interaction with other components of the NPC. Indeed, yeast Dbp5 also interacts with Gle1p (Strahm et al. 1999), an NPC-associated protein showing genetic interaction with Nup159p and also implicated in mRNA export (Del Priore et al. 1996; Murphy and Wente 1996).

Apart from interacting directly with Dbp5 via its N-terminal domain, CAN interacts with TAP and with CRM1 via its C-terminal domain, comprising multiple FG-dipeptide repeats (Katahira et al. 1999; Bachi et al. 2000; Fornerod and Ohno, this Vol.). Its interaction with Dbp5, TAP and CRM1 suggest that CAN/Nup159 may act as a platform on which the mRNP and CRM1-cargo complexes are dismantled and from which export factors can be immediately returned to the nucleus.

6 Perspectives

The past few years have seen great progress in the characterization of the mRNA export pathway and many questions have been answered, but many still remain open. It is widely accepted that export of cellular mRNPs might not be mediated by the importin-β-like family of transport receptors, and therefore that different mechanisms may account for the selectivity and directionality of this process. Indeed, one issue that certainly needs to be addressed before RAE1/Gle2p and TAP/Mex67p can be definitively promoted to the category of nuclear export receptors is the mechanism by which these proteins mediate directional transport of mRNP cargoes across the NPC. How do these proteins release their cargoes after translocation? In addition, it will be interesting to determine the role of RAE1 and of the different NXFs in metazoans. The mechanism for movement of macromolecules through the pore remains to be elucidated. An unexpected observation that sheds light on this process is that transport receptors, on their own and in association with small substrates, can cross the NPC in the absence of GTP hydrolysis. It is not yet clear if the translocation of large RNP is also energy independent or if there are GTPases or ATPases involved. A clear example of the involvement of an ATPase in mRNA export is Dbp5, which is an RNA-dependent ATPase. It has been proposed that Dbp5 plays an important role at the terminal step of the export process, but direct evidence is still missing. Moreover, Dbp5 is required for the export of mRNPs but does not play an important role in tRNA export, nor in CRM1-mediated export, raising the question of how its functional specificity is determined. Apart from the mechanism of translocation, probably the least clear aspect of RNP transport is the mechanism of nuclear retention and selective removal of retention factors before translocation. It is also currently unclear whether cellular mRNAs are exported by a unique export pathway or whether multiple export routes are available.

Acknowledgments. I wish to thank Elena Conti and Christel Schmitt for critical reading of the manuscript.

References

Alzhanova-Ericsson AT, Sun X, Visa N, Kiseleva E, Wurtz T, Daneholt B (1996) A protein of the SR family of splicing factors binds extensively to exonic Balbiani ring pre-mRNA and accompanies the RNA from the gene to the nuclear pore. Genes Dev 10:2881–2893

Bachi A, Braun IC, Rodrigues JP, Panté N, Ribbeck K, von Kobbe C, Kutay U, Wilm M, Görlich D, Carmo-Fonseca M, Izaurralde E (2000) The C-terminal domain of TAP interacts with the nuclear pore complex and promotes export of specific CTE-bearing RNA substrates. RNA 6:136–158

Bailer SM, Siniossoglou S, Podtelejnikov A, Hellwig A, Mann M, Hurt E (1998) Nup116p and Nup100p are interchangeable through a conserved motif which constitutes a docking site for the mRNA transport factor Gle2p. EMBO J 17:1107–1119

Bataillé N, Helser T, Fried HM (1990) Cytoplasmic transport of ribosomal subunits microinjected into the *Xenopus laevis* oocyte nucleus: a generalized, facilitated process. J Cell Biol 111: 1571–1582

Bear J, Tan W, Zolotukhin AS, Tabernero C, Hudson EA, Felber BK (1999) Identification of novel import and export signals of human TAP, the protein that binds to the CTE element of the type D retrovirus mRNAs. Mol Cell Biol 19:6306–6317

Belgareh N, Snay-Hodge C, Pasteau F, Dagher S, Cole C, Doye V (1998) Functional characterization of a Nup159p-containing nuclear pore subcomplex. Mol Biol Cell 9:3475–3492

Black BE, Lévesque L, Holaska JM, Wood TC, Paschal B (1999) Identification of an NTF2-related factor that binds RanGTP and regulates nuclear protein export. Mol Cell Biol 19:8616–8624

Black BE, Holaska JM, Lévesque L, Ossareh-Nazari B, Carol Gwizdek C, Dargemont C, Paschal BM (2001) NXT1 is necessary for the terminal step of CRM1-mediated nuclear export. J Cell Biol 152:141–156

Boer JM, Bonten-Surtel J, Grosveld G (1998) Overexpression of the nucleoporin CAN/Nup214 induces growth arrest, nucleocytoplasmic transport defects, and apoptosis. Mol Cell Biol 18: 1236–1247

Braun IC, Rohrbach E, Schmitt C, Izaurralde E (1999) TAP binds to the constitutive transport element (CTE) through a novel RNA-binding motif that is sufficient to promote CTE-dependent RNA export from the nucleus. EMBO J 18:1953–1965

Braun IC, Herold A, Rode M, Izaurralde E (2001) Overexpression of human TAP/NXF1 bypasses nuclear retention and stimulates mRNA nuclear export. J Biol Chem (published electronically 19 March 2001)

Brown JA, Bharathi A, Ghosh A, Whalen W, Fitzgerald E, Dhar R (1995) A mutation in the schizosaccharomyces pombe rae1 gene causes defects in poly(A)+ RNA export and in the cytoskeleton. J Biol Chem 270:7411–7419

Daneholt B (1997) A look at messenger RNP moving through the nuclear pore. Cell 88:585–588

De la Cruz J, Kressler D, Linder P (1999) Unwinding RNA in *Saccharomyces cerevisiae*: DEAD-box proteins and related families. Trends Biochem Sci 24: 192–198

Del Priore V, Snay CA, Bahr A, Cole CN (1996) The product of the Saccharomyces cerevisiae RSS1 gene, identified as a high-copy suppressor of the rat7–1 temperature-sensitive allele of the RAT7/NUP159 nucleoporin, is required for efficient mRNA export. Mol Biol Cell 7:1601–1621

Del Priore V, Heath CV, Snay-Hodge CA, MacMillan A, Gorsch LC, Dagher S, Cole CN (1997) A structure/function analysis of Rat7p/Nup159p, an essential nucleoporin of *Saccharomyces cerevisiae*. J Cell Sci 110:2987–2999

Dieckmann T, Withers-Ward ES, Jarosinski MA, Liu CF, Chen IS, Feigon J (1998) Structure of a human DNA repair protein UBA domain that interacts with HIV-1 Vpr. Nat Struct Biol 5:1042–1047

Dreyfuss G, Matunis MJ, Piñol-Roma S, Burd CG (1993) HnRNP proteins and the biogenesis of mRNA. Annu Rev Biochem 62:289–321

Flach J, Bossie M, Vogel J, Corbett A, Jinks T, Willins DA, Silver PA (1994) A yeast RNA-binding protein shuttles between nucleus and cytoplasm. Mol Cell Biol 14:8399–8407

Gabler S, Schüttt H, Groitl P, Wolf H, Shenk T, Dobner T (1998) E1B 55-kilodalton-associated protein: a cellular protein with RNA-binding activity implicated in nucleocytoplasmic transport of adenovirus and cellular mRNAs. J Virol 72:7960–7971

Gorsch LC, Dockendorff TC, Cole CN (1995) A conditional allele of a novel repeat containing yeast nucleoporin Rat7/Nup159 causes both rapid cessation of mRNA export and reversible clustering in nuclear pore complexes. J Cell Biol 4939–955

Grüter P, Tabernero C, von Kobbe C, Schmitt C, Saavedra C, Bachi A, Wilm M, Felber BK, Izaurralde E (1998) TAP, the human homologue of Mex67p, mediates CTE-dependent RNA export from the nucleus. Mol Cell 1:649–659

Guzik BW, Levesque L, Prasad S, Bor Y-C, Balck BE, Paschal BM, Rekosh D, Hammarskjold M-L (2001) NXT1/p15 is a crucial cellular cofactor in TAP dependent export of intron-containing RNA in mammalian cells. Mol Cell Biol 7:2545–2554

Hamm J, Mattaj IW (1990) Monomethylated cap structures facilitate RNA export from the nucleus. Cell 63:109–118

Herold A, Suyama M, Rodrigues JP, Braun IC, Kutay U, Carmo-Fonseca M, Bork P, Izaurralde E (2000) TAP/NXF1 belongs to a multigene family of putative RNA export factors with a conserved modular architecture. Mol Cell Biol 20:8996–9008

Hodge CA, Colot HV, Stafford P, Cole CN (1999) Rat8p/Dbp5p is a shuttling transport factor that interacts with rat7p/Nup159p and Gle1p and suppresses the mRNA export defect of xpo1-1 cells. EMBO J 18:5778–5788

Hurwitz ME, Strambio-de-Castillia C, Blobel G (1998) Two yeast nuclear pore complex proteins involved in mRNA export from cytoplasmically oriented subcomplex. Proc Natl Acad Sci USA 95:11242–11245

Izaurralde E, Lewis J, McGuigan C, Jankowska M, Darzynkiewicz E, Mattaj IW (1994) A nuclear cap binding protein complex involved in pre-mRNA splicing. Cell 78:657–668

Izaurralde E, Lewis J, Gamberi C, Jarmolowski A, McGuigan C, Mattaj IW (1995) A cap-binding protein complex mediating U snRNA export. Nature 376:709–712

Izaurralde E, Jarmolowski A, Beisel C, Mattaj IW, Dreyfuss G, Fischer U (1997) A role for the M9 transport signal of hnRNP A1 in mRNA nuclear export. J Cell Biol 137:27–35

Jarmolowski A, Boelens WC, Izaurralde E, Mattaj IW (1994) Nuclear export of different classes of RNA is mediated by specific factors. J Cell Biol 124:627–635

Kang Y, Cullen BR (1999) The human TAP protein is a nuclear mRNA export factor that contains novel RNA-binding and nucleocytoplasmic transport sequences. Genes Dev 13: 1126–1139

Kang Y, Bogerd HP, Cullen BR (2000) Analysis of cellular factors that mediate nuclear export of RNAs bearing the Mason-Pfizer monkey virus constitutive transport element. J Virol 74: 5863–5871

Katahira J, Strässer K, Podtelejnikov A, Mann M, Jung JU, Hurt E (1999) The Mex67p-mediated nuclear mRNA export pathway is conserved from yeast to human. EMBO J 18: 2593–2609

Kataoka N, Yong J, Kim VN, Velazquez F, Perkinson RA, Wang F, Dreyfuss G (2000) Pre-mRNA splicing imprints mRNA in the nucleus with a novel RNA-binding protein that persists in the cytoplasm. Mol Cell 6:673–682

Kiseleva E, Wurtz T, Visa N, Daneholt B (1994) Assembly and disassembly of spliceosomes along a specific pre-messenger RNP fiber. EMBO J 13:6052–6061

Kraemer D, Blobel G (1997) mRNA binding protein mrnp 41 localizes to both nucleus and cytoplasm. Proc Natl Acad Sci USA 94:9119–9124

Kraemer D, Wozniak RW, Blobel G, Radu A (1994) The human CAN protein, a putative oncogene product associated with myeloid leukemogenesis, is a nuclear pore complex protein that faces the cytoplasm. Proc Natl Acad Sci USA 91:1519–1523

Kraemer DM, Strambio-de-Castillia C, Blobel G, Rout MP (1995) The essential yeast nucleoporin NUP159 is located on the cytoplasmic side of the nuclear pore complex and serves in karyopherin-mediated binding of transport substrate. J Biol Chem 270:19017–19021

Kutay U, Izaurralde E, Bischoff FR, Mattaj I.W, Görlich D (1997) Dominant-negative mutants of importin-β block multiple pathways of import and export through the nuclear pore complex. EMBO J 16:1153–1163

Lee MS, Henry M, Silver PA (1996) A protein that shuttles between the nucleus and the cytoplasm is an important mediator of RNA export. Genes and Dev 10:1233–1246

Legrain P, Rosbash M (1989) Some cis- and trans-acting mutants for splicing target pre-mRNA to the cytoplasm. Cell 57:573–583

Le Hir H, Maquat EL, Moore JM (2000a) Pre-mRNA splicing alters mRNP composition: evidence for stable association of proteins at exon-exon junctions. Genes Dev 14:1098–1108

Le Hir H, Izaurralde E, Maquat LE, Moore MJ (2000b) The spliceosome deposits multiple proteins 20–24 nucleotides upstream of mRNA exon-exon junctions. EMBO J 19:6860–6869

Liker E, Fernandez E, Izaurralde E, Conti E (2000) The structure of the mRNA nuclear export factor TAP reveals a cis arrangement of a non-canonical RNP domain and an LRR domain. EMBO J 19:5587–5598

Lindsay S, Shopland, Lawrence JB (2000) Seeking common ground in nuclear complexity. J Cell Biol 150:1–4

Luo M-J, Reed R (1999) Splicing is required for rapid and efficient mRNA export in metazoans. Proc Natl Acad Sci USA 96:14937–14942

McGarvey T, Rosonina E, McCracken S, Li Q, Arnaout R, Mientjes E, Nickerson AJ, Awrey D, Greenblatt J, Grosveld G, Blencowe BJ (2000) The acute myeloid leukemia-associated protein DEK forms a splicing-dependent intercation with exon-product complexes. J Cell Biol 150:309–320

Mehlin H, Daneholt B, Skoglund U (1992) Translocation of a specific premessenger ribonucleoprotein particle through the nuclear pore complex studied with electron microscope tomography. Cell 69:605–613

Misteli T, Spector DL (1998) The cellular organization of gene expression. Curr Opin Cell Biol 10:323–331

Murphy R, Wente SR (1996) An RNA-export mediator with an essential nuclear export signal. Nature 383:357–360

Murphy R, Watkins JL, Wente SR (1996) GLE2, a *Saccharomyces cerevisiae* homologue of the *Schizosaccharomyces pombe* export factor RAE1, is required for nuclear pore complex structure and function. Mol Biol Cell 7:1921–1937

Pasquinelli AE, Ernst RK, Lund E, Grimm C, Zapp ML, Rekosh D, Hammarskjold, M-L, Dahlberg JE (1997) The constitutive transport element (CTE) of Mason-Pfizer Monkey Virus (MPMV) accesses an RNA export pathway utilized by cellular messenger RNAs. EMBO J 16:7500–7510

Pokrywka NJ, Goldfarb DS (1995) Nuclear export pathways of tRNA and 40 S ribosomes include both common and specific intermediates. J Biol Chem 270:3619–3624

Portman DS, O'Connor P Dreyfuss G (1997) YRA1, an essential *Saccharomyces cerevisiae* gene, encodes a novel nuclear protein with RNA annealing activity. RNA 3:527–537

Pritchard CEJ, Fornerod M, Kasper JH, van Deursen JMA (1999) RAE1 is a shuttling mRNA export factor that binds to a GLEBS-like NUP98 motif at the nuclear pore complex through multiple domains. J Cell Biol 145:237–253

Rodrigues JP, Rode M, Gatfield D, Blencowe BJ, Carmo-Fonseca M, Izaurralde E (2001) REF proteins mediate the export of spliced and unspliced mRNAs from the nucleus. Proc Natl Acad Sci USA 98:1030–1035

Russell I, Tollervey D (1995) Yeast Nop3p has structural and functional similarities to mammalian pre-mRNA binding proteins. Eur J Cell Biol 66:293–301

Rutz B, Seraphin B (2000) A dual role for BBP/ScSF1 in nuclear pre-mRNA retention and splicing. EMBO J 19:1873–1886

Saavedra CA, Felber BK, Izaurralde E (1997) The simian retrovirus-1 constitutive transport element CTE, unlike HIV-1 RRE, utilises factors required for cellular RNA export. Curr Biol 7:619–628

Santos-Rosa H, Moreno H, Simos G, Segref A, Fahrenkrog B, Panté N, Hurt E (1998) Nuclear mRNA export requires complex formation between Mex67p and Mtr2p at the nuclear pores. Mol Cell Biol 18:6826–6838

Sabri N, Visa N (2000) The Ct-RAE1 protein interacts with Balbiani ring RNP particles at the nuclear pore. RNA 6:1597–1609

Schmitt C, von Kobbe C, Bachi A, Panté N, Rodrigues JP, Boscheron C, Rigaut G, Wilm M, Séraphin B, Carmo-Fonseca M, Izaurralde E (1999) Dbp5, a DEAD box-protein required for mRNA export, is recruited to the cytoplasmic fibrils of nuclear pore complex via a conserved interaction with CAN/Nup159p. EMBO J 18:4332–4347

Segref A, Sharma K, Doye V, Hellwig A, Huber J, Lührmann R, Hurt E (1997) Mex67p, a novel factor for nuclear mRNA rexport binds to both poly(A)+ RNA and nuclear pores. EMBO J 16:3256–3271

Singleton DR, Shaoping C, Hitomi C, Kumagai C, Tartakoff A (1995) A yeast protein that bidirectionally affects nucleocytoplasmic transport. J Cell Sci 108:265–272

Snay-Hodge C, Colot H, Goldstein AL, Cole CN (1998) Dbp5p/Rat8p is a yeast nuclear pore-associated DEAD-box protein essential for RNA export. EMBO J 17:2663–2676

Sträßer K, Hurt E (2000) Yra1p, a conserved nuclear RNA-binding protein, interacts directly with Mex67p and is required for mRNA export. EMBO J 19:410–420

Sträßer K, Baßler J, Hurt E (2000) Binding of the Mex67p/Mtr2p heterodimer to FXFG, GLFG, and FG repeat nucleoporins is essential for nuclear mRNA export. J Cell Biol 150:695–706

Strahm Y, Fahrenkrog B, Zenklusen D, Rychner E, Kantor J, Rosbash M, Stutz F (1999) The RNA export factor Gle1p is located on the cytoplasmic fibrils of the NPC and physically interacts with the FG-nucleoporin Rip1p, the DEAD-box protein Rat8p/Dbp5p and a new protein Ymr255p. EMBO J 18:5761–5777

Strawn LA, Shen T, Wente SR (2000) The GLFG region of Nup116p and Nup100p serve as binding sites for both Kap95p and Mex67p at the nuclear pore complex. J Biol Chem 276:6445–6552

Stutz F, Bachi A, Doerks T, Braun IC, Séraphin B, Wilm M, Bork P, Izaurralde E (2000) REF, an evolutionarily conserved family of hnRNP-like proteins, interacts with TAP/Mex67p and participates in mRNA nuclear export. RNA 6:638–650

Suyama M, Doerks T, Braun IC, Sattler M, Izaurralde I, Bork P (2000) Prediction of structural domains of TAP reveals details of its interaction with p15 and nucleoporins. EMBO Rep 1:53–58

Tan W, Zolotukhin AS, Bear J, Patenaude DJ, Felber BK (2000) The mRNA export in *C. elegans* is mediated by Ce-NXF-1, an ortholog of human TAP/NXF1 and *S. cerevisiae* Mex67p. RNA 6: 1762–1772

Tseng SS-I, Weaver PL, Hitomi M, Tartakoff AM, Chang T-H (1998) Dbp5p, a cytosolic RNA helicase, is required for poly(A)+ RNA export. EMBO J 17:2651–2662.

Van Deursen J, Boer J, Kasper L, Grosveld G (1996) G2 arrest and impaired nucleocytoplasmic transport in mouse embryos lacking the proto-oncogene CAN/Nup214. EMBO J 15:5574–5583

Visa N, Izaurrralde E, Ferreira J, Daneholt B, Mattaj IW (1996a) A nuclear cap-binding complex binds balbiani ring pre-mRNA cotranscriptionally and accompanies the ribonucleoprotein particle during nuclear export. J Cell Biol 133:4–14

Visa N, Alzhanova-Ericsson AT, Sun X, Kiseleva E, Björkroth B, Wurtz T, Daneholt B (1996b) A pre-mRNA-binding protein accompanies the RNA from the gene through the nuclear pores into polysomes. Cell 84:253–264

Wilson SM, Datar KV, Paddy MR, Swedlow JR, Swanson M (1994) Characterization of nuclear polyadenylated RNA-binding proteins in *Saccharomyces cerevisiae.* J Cell Biol 127:1173–1184

Wu X, Kasper LH, Mantcheva RT, Mantchev GT, Springett MJ, van Deursen JM (2001) Disruption of the FG nucleoporin NUP98 causes selective changes in nuclear pore complex stoichiometry and function. Proc Natl Acad Sci USA 98:3191–3196

Yoon JH, Whalen WA, Bharathi A, Shen R, Dhar R (1997) Npp106p, a *Schizosaccharomyces pombe* nucleoporin similar to *Saccharomyces cerevisiae* Nic96p, functionally interacts with Rae1p in mRNA export. Mol Cell Biol 17:7047–7060

Yoon JH, Love DC, Guhathakurta A, Hanover JA, Dhar R (2000) Mex67p of *Schizosaccharomyces pombe* interacts with Rae1p in mediating mRNA export. Mol Cell Biol 20:8767–8782

Zhou Z, Luo M-J, Strasser K, Katahira J, Hurt E, Reed R (2000) The protein Aly links pre-messenger-RNA splicing to nuclear export in metazoans. Nature 407:401–405

Using Retroviruses To Study the Nuclear Export of mRNA

Bryan R. Cullen[1]

1 Nuclear mRNA Export and the Retroviral Life Cycle

Several non-exclusive approaches can be taken to identify and then characterize cellular proteins involved in the nucleocytoplasmic transport of protein or RNA molecules. One approach that has been extremely useful is the genetic identification of factors required for the nucleocytoplasmic transport of a particular macromolecule, e.g. nuclear export of poly(A)$^+$ RNA, by screening yeast conditional mutants for, in this case, nuclear accumulation of poly(A)$^+$ RNA at the restrictive temperature. While this kind of approach has led to the identification of gene products that play a direct or indirect role in several transport pathways, including mRNA export (Amberg et al. 1992; Kadowaki et al. 1994), it can be difficult to subsequently discern the biochemical role played by a particular genetically identified gene product. An alternative strategy is to use protein binding affinity, either in vitro or using the yeast two-hybrid protein:protein interaction assay, to identify factors that specifically bind to a relevant protein target. For example, the G-protein Ran, in its GTP-βound form, is a critical cofactor for nuclear transport mediated by all members of the extensive importin-β/karyopherin-β family of transport factors (Melchior et al. 1993; Moore and Blobel 1993; Izaurralde et al. 1997). The identification of human proteins that can bind to the GTP-, but not GDP-, bound form of Ran in vitro has therefore facilitated the identification of several nuclear transport factors, including CAS and exportin t (Kutay et al. 1997, 1998). Conversely, use of the two-hybrid assay allowed the initial identification of transportin/karyopherin-β2, the import factor specific for the unusual type of nuclear localization signal (NLS) present in the heterogeneous nuclear ribonucleoprotein A1 (hnRNP A1; Pollard et al. 1996; Fridell et al. 1997).

A third approach that has been particularly useful in the identification of cellular nuclear export factors relies on an unusual idiosyncrasy of the retroviral replication cycle. Retroviruses are unique among viruses in that the viral genome is an mRNA molecule that undergoes reverse transcription, to form a double-stranded DNA provirus, as an obligate step in the replication cycle (reviewed by Varmus and Brown 1989). The DNA provirus integrates into the

[1] Howard Hughes Medical Institute and Department of Genetics, Room 426 CARL Building, Research Drive, Durham, NC 27710, USA

Results and Problems in Cell Differentiation, Vol. 35
K. Weis (Ed.): Nuclear Transport

genomic DNA of the target cell and, in the case of most retroviruses, is then transcribed to give rise to a single initial genome-length viral mRNA transcript. This initial transcript can then be spliced by the host RNA processing machinery, using one or more 5′ and 3′ splice sites, to generate singly and, for some viruses, also multiply spliced mRNAs. Importantly, not only these processed viral mRNAs but also the initial unspliced transcript must be exported to the cytoplasm and then translated and, in the case of the unspliced viral mRNA, also packaged into progeny virions for the viral life cycle to proceed.

One unusual aspect of the retroviral life cycle therefore lies in the fact that these viruses generally encode a single RNA transcript that must be exported from the nucleus both in an unspliced form and in one or more spliced forms. However, splicing of retroviral RNAs is entirely mediated by cellular factors that bind to 5′ and 3′ splice sites and flanking RNA sequences such as the branch point and the polypyrimidine tract. Binding of a subset of these factors not only commits these RNAs to the splicing pathway, but also prevents incompletely spliced mRNAs from leaving the nucleus by blocking access to the cellular mRNA export pathway (Chang and Sharp 1989; Legrain and Rosbash 1989). This nuclear retention mechanism, which is probably designed to prevent the cytoplasmic translation of intron-containing cellular pre-mRNAs that would be likely to encode useless or even deleterious proteins, probably constitutes a quality control mechanism. However, these cellular "commitment factors" present retroviruses with a serious problem. On the one hand, retroviruses are dependent on the host cell to generate the essential spliced forms of their initial transcript and they must therefore carry recognizable splice sites. On the other hand, the recognition of these same splice sites by commitment factors has the potential to block the nuclear export of this initial transcript, a step that is just as critical for retroviral replication. Retroviruses have therefore had to evolve mechanisms to target their unspliced, genomic mRNA into alternate nuclear export pathways that are not inhibited by commitment factor recruitment. It is now clear that these novel mRNA export pathways are largely or entirely distinct from the export pathway used by mature cellular or viral mRNAs.

2 Crm1-Dependent Retroviral mRNA Export

2.1 The Human Immunodeficiency Virus Rev Protein

Retroviruses can be broadly subdivided into a simple and a complex class (Table 1). Simple retroviruses, such as avian leukemia virus (ALV, Fig. 1) or murine leukemia virus (MLV), encode only three or, maximally, four genes, none of which is a transcription factor. The three genes common to all retroviruses are *gag*, which encodes the viral structural proteins; *pol*, which encodes the various enzymes that are critical to the viral life cycle and *env*, which encodes the viral envelope glycoprotein (Fig. 1). Simple retroviruses such as

Table 1. Major taxonomic divisions among retroviruses

Category and Subgroup[a]	Prototype[b]	Other examples[b]	Export pathway
Complex retroviruses			
Lentiviruses	HIV-1	HIV-2, SIV, VMV, FIV, EIAV	RRE/Rev/Crm1
T-cell leukemia viruses	HTLV-I	HTLV-II, STLV, BLV	RxRE/Rex/Crm1
Foamy viruses	HFV	SFV, BFV	?
Simple retroviruses			
C-type retroviruses group A	ALV	ASV, RSV	CTE/?
C-type retroviruses group B	MLV	FeLV, MSV, SNV, SSV	?
B-type retroviruses	MMTV		?
D-type retroviruses	MPMV	SRV-1	CTE/Tap
Human endogenous retroviruses	HERV-K		K-RRE/K-Rev/Crm1

[a] While all seven exogenous virus families are shown, only one human endogenous virus family, out of the 22 known, is listed.

[b] SIV, Simian immunodeficiency virus; *VMV*, visna maedi virus; *STLV*, simian T-cell leukemia virus; *BLV*, bovine leukemia virus; *HFV*, human foamy virus; *BFV* bovine foamy virus; *ASV* avian sarcoma virus; *RSV*, Rous sarcoma virus; *FeLV*, feline leukemia virus; *MSV*, murine sarcoma virus; *SNV*, spleen necrosis virus; *SSV*, simian sarcoma virus; *MMTV*, mouse mammary tumor virus; *SRV-1*, simian retrovirus type 1.

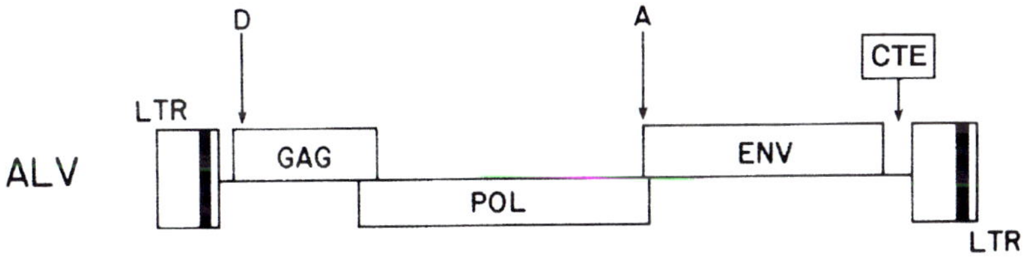

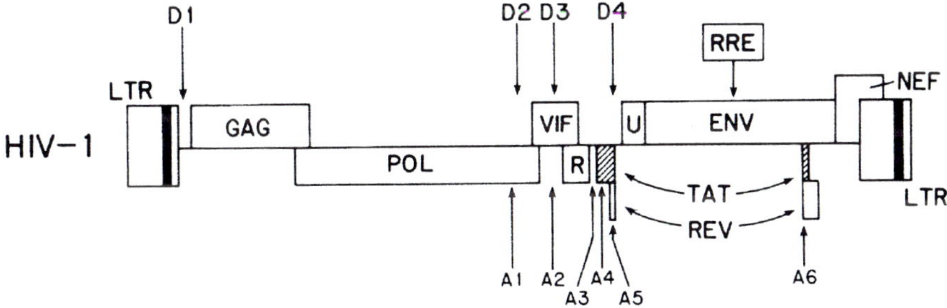

Fig. 1. Comparison of the genomic organization of the simple retrovirus ALV and the complex retrovirus HIV-1. Viral genes are indicated, as are 5′ splice sites (*D*) and 3′ splice sites (*A*). The localization in HIV-1 of the RRE RNA target for Rev is shown, as is the location of the ALV CTE

ALV encode only two viral RNAs, i.e. the initial, genome-length transcript and a singly spliced mRNA that encodes Env. As we will discuss later, nuclear export of the incompletely spliced RNA in ALV is dependent on a cellular factor that remains unidentified.

In contrast to ALV, the pathogenic retrovirus human immunodeficiency virus type 1 (HIV-1) is the prototype of the complex class of retroviruses. As shown in Fig. 1, HIV-1 encodes a large number of splice sites, and the initial genome-length HIV-1 transcript can be spliced to give at least five singly spliced mRNAs and >15 multiply or fully spliced mRNAs. Further, HIV-1 encodes not only *gag*, *pol* and *env* but also six additional genes that have no equivalent in ALV (Fig. 1) and considerable effort has been expended over the years to identify the role and mechanism of action of each of these novel retroviral auxiliary proteins. Relevant here is the finding that mutational inactivation of the Rev open reading frame has no effect on the expression of the various fully spliced viral mRNAs but entirely blocks the expression of the unspliced or singly spliced viral transcripts (Feinberg et al. 1986; Malim et al. 1988). While it was initially suggested that Rev might act as a selective inhibitor of viral RNA splicing (Feinberg et al. 1986), it soon became apparent that the pattern of viral RNA expression in the nucleus of transfected cells was not significantly changed in the presence or absence of Rev (Malim et al. 1989b). Rather, it was observed was that the incompletely spliced viral RNAs were unable to leave the nucleus in the absence of Rev but were efficiently exported when the *rev* gene remained intact. This effect was also shown to require a *cis*-acting RNA target site located within the viral *env* gene, the Rev response element (RRE), that forms a complex ~234-nucleotide RNA secondary structure (Fig. 1; Malim et al. 1989b). On the basis of these data, it was therefore proposed that Rev functions as a viral RNA sequence-specific nuclear RNA export factor (Malim et al. 1989b), a hypothesis that has been subsequently confirmed by data from a range of different systems and laboratories.

In fact, it is now clear that Rev, which is one of three early gene products encoded by the fully spliced HIV-1 mRNAs (the others are Tat, a transcription factor, and Nef, which enhances HIV-1 progeny virion release and infectivity), serves as the molecular switch that activates the expression of the late, mostly structural, gene products that are encoded by the incompletely spliced viral mRNAs. While the fully spliced viral mRNAs encoding the viral early proteins Tat, Rev and Nef are exported via the canonical cellular mRNA export pathway, Rev is required to selectively activate the nuclear export, and hence translation, of these incompletely spliced mRNAs. Therefore, in the absence of functional Rev, only the three early viral proteins are expressed and viral replication is aborted.

Efforts to further characterize the mechanism of action of Rev identified an arginine-rich RNA binding motif (ARM) that mediates high-affinity binding to a single site located within the RRE (Fig. 2; Heaphy et al. 1990; Bartel et al. 1991; Malim and Cullen 1991). This arginine motif also serves as the Rev NLS and was subsequently shown to serve as the direct binding site for

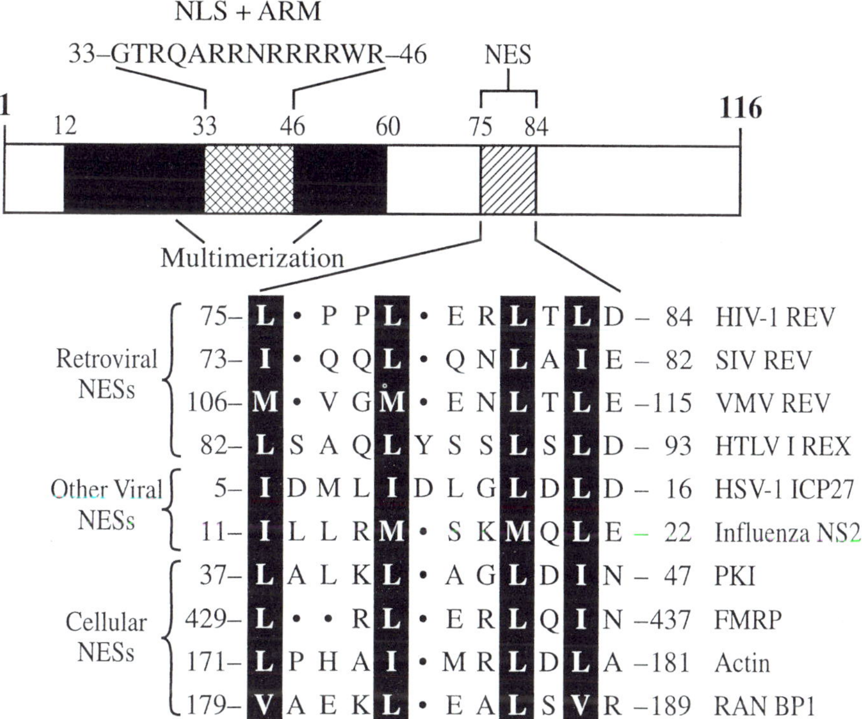

Fig. 2. Functional organization of the HIV-1 Rev protein. The location and sequence of the Rev basic domain and leucine-rich NES are delineated. In addition, the the sequence of several other known leucine-rich NESs, many of which have been demonstrated to functionally substitute for the Rev NES in mediating unspliced RNA export, is shown. *HSV-1* herpes simplex virus type 1, *PKI* protein kinase inhibitor, *FMRP* fragile X mental retardation protein. (Cullen 1998)

the importin-β nuclear import factor (Malim et al. 1989a; Henderson and Percipalle 1997; Truant and Cullen 1999). Closely flanking the ARM lie the multimerization domains of Rev. These essential sequences promote the assembly of multiple Rev molecules onto the RRE in a process mediated by specific protein:protein and apparently relatively nonspecific protein:RNA interactions (Malim and Cullen 1991; Zapp et al. 1991). A third and final functional domain located within the 116-amino-acid Rev protein is a 10-amino-acid leucine-rich motif (NH_2-LPPLERLTLD-COOH) that is required for Rev function but not for RRE binding or multimerization (Fig. 2; Malim et al. 1989a, 1991). As a result, mutant Rev proteins lacking a functional form of this motif compete with wild-type Rev for binding to the RRE and can inhibit the Rev-induced activation of viral late gene expression.

Further efforts to define the functional role of the Rev leucine motif led to the demonstration that this sequence is a nuclear export signal (NES), i.e. that a protein bearing this sequence in a functional form is efficiently exported from the nucleus (Fischer et al. 1995; Wen et al. 1995). Rev is therefore a nucleocytoplasmic shuttle protein that contains an NLS coincident with its arginine motif and an NES coincident with its leucine motif (Fig. 2; Meyer and Malim 1994). However, mutational inactivation of the Rev leucine motif results in a Rev protein that remains sequestered in the cell nucleus.

Previous work had demonstrated that different classes of cellular RNA, e.g. mRNA, tRNA and U snRNA (U-rich small nuclear RNA), could competitively inhibit their own export from microinjected *Xenopus* oocyte nuclei at concentrations that had no effect on the export of heterologous RNAs (Jarmolowski et al. 1994). It was therefore possible to ask whether Rev-dependent RNA export occurred via the cellular mRNA export pathway or via some other mechanism. To examine this question, Fischer et al. (1995) injected saturating levels of the Rev NES into the nucleus of *Xenopus* oocytes and asked which RNA export pathways would be affected. Under these conditions, the authors observed effective inhibition of not only Rev-dependent RNA export but also of 5S rRNA and U1 snRNA export. In contrast, mRNA and tRNA export remained unaffected. Subsequently, evidence was presented arguing that both 5S rRNA and U snRNA export is mediated by cellular proteins that contain leucine-rich NESs which are critical for their RNA export function (Fridell et al. 1996; Ohno et al. 2000). In fact, leucine-rich NESs functionally equivalent to that seen in Rev have now been defined on a large number of cellular proteins, many of which clearly have nothing to do with RNA export (Fig. 2).

Efforts to identify the cellular target for the Rev NES, using both biochemical and genetic approaches, led to the demonstration that the Rev leucine-rich NES is a specific binding site for the cellular protein Crm1 (Fornerod et al. 1997a; Stade et al. 1997). Crm1 is a member of the importin-β/karyopherin-β family of nucleocytoplasmic transport factors that are characterized by their affinity for Ran in its GTP-βound form (reviewed by Görlich and Kutay 1999). In fact, Ran·GTP is a critical cofactor for Rev binding by Crm1 and hydrolysis of Ran·GTP to Ran·GDP results in release of Crm1 from the Rev protein (Fornerod et al. 1997a; Bogerd et al. 1998). Because this hydrolysis requires both the Ran GTPase activating protein RanGAP and a cellular cofactor called Ran binding protein 1 (RanBP1), both of which are largely confined to the cytoplasm, this hydrolysis step provides a mechanism to promote the release of Rev, and its bound RNA cargo, after export to the cytoplasm (Izaurralde et al. 1997). How the export of the large RRE:Rev:Crm1:Ran·GTP complex is actually carried out remains unclear. However, it is known that Crm1 can directly interact with several components of the nuclear pore complex (NPC) including the nucleoporin Nup214/CAN (Fornerod et al. 1997b; Neville et al. 1997). Importantly, selective inhibitors of Crm1 function, such as dominant negative forms of Nup214/CAN or the drug leptomycin B, rapidly and specifically block Rev-dependent RNA export but have little effect on cellular or early HIV-1 mRNA export (Fornerod et al. 1997a; Bogerd et al. 1998). Therefore, these data imply that Crm1 is not directly required for mRNA export from the nucleus.

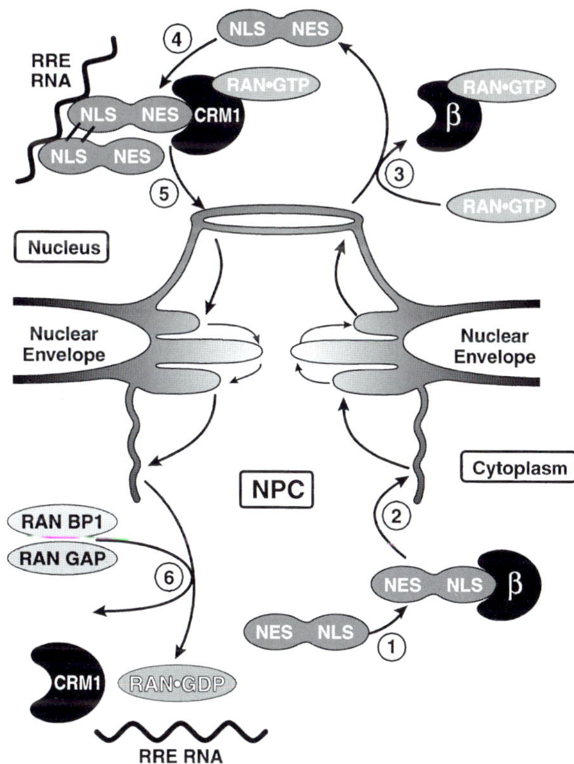

Fig. 3. Overview of HIV-1 Rev-dependent nuclear RNA export and nucleocytoplasmic shuttling. The steps involved in these processes are described in the text

In conclusion, the mechanism of action of Rev can be summarized as shown in Fig. 3. After its synthesis in the cytoplasm, Rev directly binds to importin-β via its arginine-rich NLS (1). Importin-β targets the resultant heterodimer to the NPC and then mediates its import into the nucleus in a process that likely involves binding and release of specific nucleoporins (2). Once in the nucleus, importin-β binds to Ran·GTP, resulting in the release of Rev (3) which is then free to bind to, and multimerize on, the RRE found in incompletely spliced HIV-1 transcripts (4). Rev also binds, via its leucine-rich NES, to Crm1 in its Ran·GTP bound form to generate a ribonucleoprotein complex consisting of the HIV-1 RNA bound to multiple Rev, Crm1 and Ran·GTP molecules (4). This is then targeted to the NPC, through the action of Crm1, and transported to the cytoplasm (5). There, the combined action of RanGAP and RanBP1 results in the hydrolysis of Ran·GTP to Ran·GDP, leading to dissolution of the complex (6). Rev, as well as Crm1 and Ran·GDP, are then free to reenter the nucleus, while the exported HIV-1 RNA can be used for translation of viral proteins or incorporated into progeny virions (Fig. 3).

In addition to HIV-1, all other members of the lentivirus subgroup of retroviruses also encode Rev-like proteins that are required for expression of the viral late, structural gene products. The mechanism of action of these other lentiviral Rev proteins appears very similar to that of HIV-1 and all require the Crm1 cofactor for their function. One interesting observation, however, is that only a subset of lentiviral Rev proteins contain a leucine motif comparable to the 10-amino-acid NES seen in HIV-1 Rev and in the majority of cellular proteins that are known to be exported from the nucleus in a Crm1-dependent manner (Fig. 2). In contrast, some other lentiviral Rev proteins, including equine infectious anemia virus (EIAV) Rev and feline immunodeficiency virus (FIV) Rev, bind to Crm1 via a discrete ~24-amino-acid NES motif that is clearly distinct in primary sequence from the HIV-1 Rev NES (Fridell et al. 1993; Mancuso et al. 1994). While it is likely that cellular nucleocytoplasmic shuttle proteins exist that also feature examples of this longer type of Crm1 binding sequence, none have been identified thus far.

2.2 The Human T-Cell Leukemia Virus Rex Protein

Human T cell leukemia virus type 1 (HTLV-I) is the prototype of a second family of complex retroviruses, evolutionarily distinct from HIV-1 and the other lentiviruses, that encode not only *gag*, *pol* and *env* but also at least two auxiliary proteins termed Tax and Rex (Table 1). Tax, a transcription factor and Rex, the HTLV-I homologue of the HIV-1 Rev protein, are both translated from a single fully spliced mRNA that does not require Rex for its nuclear export. In contrast, the unspliced HTLV-I *gag–pol* mRNA and the singly spliced *env* mRNA are totally dependent on Rex for their nuclear export and expression (Hidaka et al. 1988; Hanly et al. 1989).

The ~27-kDa Rex protein has been the subject of considerable scientific interest due to its critical importance for replication of the pathogenic HTLV-I. In fact, these efforts have revealed that the role and mechanism of action of Rex are very closely comparable to those of the HIV-1 Rev protein. Like Rev, Rex contains an arginine motif that serves as both an importin-β-dependent NLS and as an RNA binding motif (Grassmann et al. 1991; Palmeri and Malim 1999). Also, Rex contains a 12-amino-acid leucine-rich NES that binds to Crm1 (Fig. 2), as well as sequences required for Rex multimerization (Hope et al. 1991; Weichselbraun et al. 1992; Hakata et al. 1998). However, the Rex protein, which is considerably larger than HIV-1 Rev, has no obvious sequence homology to Rev and these functional domains are clearly arranged in a different order in Rev and Rex. While the Rex protein has a different RNA sequence specificity than Rev, Rex also binds to a highly structured viral RNA target element, the Rex response element (RxRE; Bogerd et al. 1991). Interestingly, the ~279-nucleotide RxRE differs from the ~234-nucleotide RRE in that the former is located within the U3 and R regions of the HTLV-I long terminal repeat, not in the viral *env* gene (Fig. 1; Hanly et al. 1989).

2.3 The Human Endogenous Retrovirus K K-Rev Protein

The lentiviruses, including HIV-1, and the T-cell leukemia viruses, of which the prototype is HTLV-I, are the only known exogenous virus subgroups that encode a Crm1-dependent nuclear RNA export factor (Table 1). Surprisingly, however, an equivalent viral protein, termed K-Rev, has recently been identified in the human endogenous retrovirus K (HERV-K) family (Magin et al. 1999; Yang et al. 1999). Endogenous retroviruses are proviruses that form part of the genome of a species (reviewed by Patience et al. 1997). They originate as exogenous retroviruses and enter the species genome by infection of a germ cell that then goes on to give rise to a progeny animal that participates in future species evolution. Because endogenous viruses are no longer subjected to positive selection for integrity of their encoded gene products, and may in fact be subjected to negative selection if they are in some way deleterious to their host, the large majority of endogenous viruses have acquired one or more inactivating mutations over time, many of which are gross deletions. As most human endogenous viruses have been part of the primate genome for millions of years, they have had ample opportunity to acquire such mutations.

The human genome consists of ~0.1% endogenous viruses and these can be divided into 22 independently acquired sub-families, one of which is the HERV-K grouping (Tristem 2000). The HERV-Ks entered the human genome starting ~30 million years ago, with the most recent integrations occurring ~5 million years ago (Medstrand and Mager 1998). There are between 50 and 170 HERV-K proviruses per haploid human genome and while many of these are obviously defective, others do not harbor any evident mutations (Tristem 2000). In fact, the HERV-Ks are probably the single most intact family of human endogenous viruses and certain HERV-K proviruses have been shown to encode functional forms of essential retroviral enzymes, such as reverse transcriptase and integrase (Berkhout et al. 1999).

While the HERV-Ks are not expressed in most normal tissues, their transcription is readily detectable in certain tumors, including seminomas and teratocarcinomas. This observation allowed Löwer et al. (1995) to characterize the mRNA molecules encoded by HERV-K proviruses. Surprisingly, HERV-K proved able to encode not only an unspliced *gag-pol* mRNA and a singly spliced *env* mRNA but also a single fully spliced mRNA that had the potential to encode a small protein of ~105 amino acids, now termed K-Rev. Subsequently, K-Rev has been shown to interact with Crm1 in a Ran·GTP-dependent manner, to multimerize and to bind to a *cis*-acting RNA target site, the K-RRE, located in the HERV-K LTR U3 region (Magin et al. 1999; Yang et al. 1999). K-Rev was also shown to activate the nuclear export of an unspliced mRNA, but only if the K-RRE was present in *cis* and Crm1 function was not inhibited. Therefore, it is apparent that the HERV-Ks represent a third family of retroviruses that encode a Crm1-dependent retroviral mRNA export factor (Table 1). An important difference is, of course, that while HIV-1 and HTLV-I are exogenous viruses that are very much alive today, the HERV-Ks are a family of ancient human endogenous viruses that are likely extinct in their exogenous form. However, the identification in an ~30-million-year-old family of endogenous viruses of an RNA export activity that is very closely related to HIV-1 Rev does imply that this mechanism likely evolved fairly early in retroviral evolution. However, as K-Rev actually shows little or no sequence homology to either HIV-1 Rev or HTLV-I Rex, the evolutionary origin of these three functionally equivalent retroviral export factors currently remains unclear.

3 Tap-Dependent Retroviral mRNA Export

While the nuclear export of the incompletely spliced mRNAs encoded by the complex retroviruses HIV-1 and HTLV-I requires a virally encoded adaptor protein able to recruit the Crm1 export factor to target RNAs, this mechanism cannot apply to simple retroviruses as these clearly do not encode any equivalent viral protein (Fig. 1). Nevertheless, simple retroviruses are faced with the same problem of how to export both spliced and unspliced forms of their initial transcript in the face of nuclear retention of intron-containing RNAs by host cell factors. One potential solution to this problem was first discovered in Mason-Pfizer monkey virus (MPMV), a simple retrovirus that forms the prototype of the type D family of simian retroviruses (Table 1). Specifically, MPMV was found to encode a *cis*-acting RNA sequence, the constitutive transport element (CTE), that not only induces the nuclear export of unspliced mRNA molecules in the absence of any MPMV gene product but also functionally substitutes for both Rev and the RRE in activating the expression of the late gene products encoded by HIV-1 (Bray et al. 1994). The CTE coincides with a 154-nucleotide RNA sequence, located between the *env* gene and the 3′ LTR, that has been shown to fold into an extensive stem-loop structure (Tabernero et al. 1996; Ernst et al. 1997). However, unlike the RRE, the MPMV CTE must obviously directly recruit a cellular, rather than a viral, factor as the first step in viral RNA export.

An important initial question was whether the CTE also functioned through recruitment of Crm1. Two lines of evidence showed this was not the case. Thus, reagents that block Crm1, and hence Rev, function were found to have no effect on CTE-dependent mRNA export (Bogerd et al. 1998). Further, the microinjection of saturating levels of the CTE into the *Xenopus* oocyte nucleus effectively inhibited both CTE-dependent RNA export and mRNA export but did not affect Rev-dependent RNA export or export of tRNA, U1 snRNA or 5S rRNA (Pasquinelli et al. 1997; Saavedra et al. 1997). Clearly, therefore, the cofactor required for MPMV CTE function was not Crm1 and was instead in some way important for the nuclear export of mature cellular mRNA molecules.

The initial identification of the CTE cofactor, termed Tap, was achieved by the biochemical purification of a cellular factor able to bind the wild-type CTE, but not to inactive CTE mutants (Grüter et al. 1998). Importantly, recombinant Tap also proved able to enhance CTE-dependent RNA export after microin-

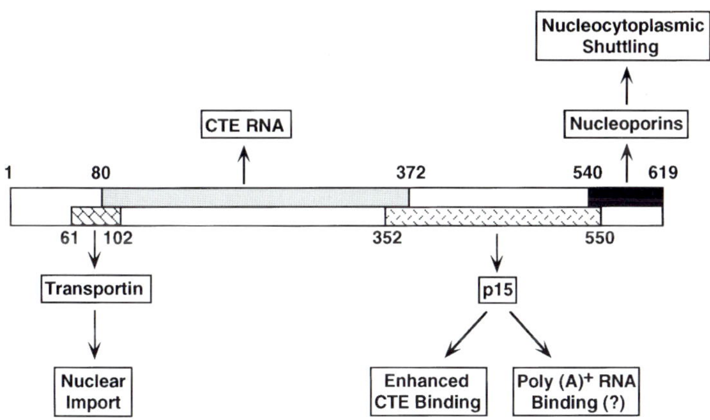

Fig. 4. Domain organization of the human Tap nuclear RNA export factor. The biological targets for several domains within Tap and the phenotypic consequence of the interaction are shown. It should be noted that the sequences required for Tap binding to cellular poly(A)$^+$ RNA have not been defined and these may be distinct from the sequences that mediate CTE binding. (Kang et al 2000)

jection into the *Xenopus* oocyte nucleus. More conveniently, human Tap was subsequently shown to rescue MPMV CTE function in an otherwise non-permissive quail cell line, thus permitting the rapid identification of sequences within Tap that were required for CTE-dependent RNA export (Kang and Cullen 1999). Functional domains identified in the 619-amino-acid Tap protein included a novel type of RNA binding domain specific for the MPMV CTE (amino acids 80–372; Braun et al. 1999; Kang and Cullen 1999), an unusual transportin-dependent NLS (amino acids 61–102; Truant et al. 1999) and most intriguingly, an essential carboxy-terminal sequence (amino acids 540–619) that proved able to mediate both nuclear import and nuclear export, i.e. that conferred on Tap the ability to shuttle between the nucleus and the cytoplasm (Fig. 4; Kang and Cullen 1999).

At this point it remained unclear whether Tap, which is not a member of the importin-β/karyopherin-β family of nucleocytoplasmic transport factors, was acting as an adaptor that recruited such a family member, i.e. in a manner comparable to Rev, or whether Tap was directly involved in mediating CTE export from the nucleus. This question was answered by the identification of cellular factors that might be required for Tap function using the yeast two-hybrid screen followed by in vitro assays. These experiments revealed that the carboxy-terminal domain of Tap could directly bind to several cellular nucleoporins including Nup214/CAN, CG1 and Nup153 (Katahira et al. 1999; Bachi et al. 2000). Importantly, extensive mutational analysis has subsequently demonstrated that the abilities of this Tap domain to bind to nucleoporins and to mediate nucleocytoplasmic shuttling are inseparable (Kang et al. 2000), thus arguing that Tap directly targets the CTE to the NPC, and hence to the cytoplasm, by binding to nucleoporins (Fig. 4).

In addition to the interaction with nucleoporins, this same screen also revealed that the central part of Tap (amino acids 352–550) could interact with a cellular cofactor termed p15 (Katahira et al. 1999). However, mutational inactivation of this interaction did not block the ability of Tap to support CTE-dependent RNA export, although data have been presented arguing that p15 can enhance CTE binding by Tap and can readily form a ternary complex with Tap and the CTE in vitro (Bachi et al. 2000; Kang et al. 2000). While not essential for CTE-dependent mRNA export, p15 may nevertheless play a critical role in cellular mRNA transport, for example by facilitating the sequence-nonspecific recruitment of Tap to cellular mRNAs (see below; Fig. 4).

At about the same time that Tap was proposed as the human cofactor for MPMV CTE-dependent RNA export, Hurt and coworkers identified a yeast gene product, Mex67p, that is critical for global mRNA export in yeast cells and that bears significant homology to human Tap (Segref et al. 1997). Mex67p was shown to interact with yeast NPCs and with a second yeast protein already known to be critical for mRNA export, termed Mtr2p (Santos-Rosa et al. 1998). While both Mex67p and Mtr2p are critical for yeast viability, they could be at least partly functionally substituted by a combination of the human Tap and p15 proteins, although Tap alone could not functionally replace Mex67p (Katahira et al. 1999). These data imply that Tap and Mex67p, and p15 and Mtr2p, are functional orthologues, although the latter pair do not show any significant sequence homology. More importantly, these data imply that both Tap and p15 are critical participants in a cellular mRNA export pathway that has been conserved from yeast to humans. The observation, noted above, that microinjection of the CTE into *Xenopus* oocytes can selectively inhibit cellular mRNA export (Pasquinelli et al. 1997; Saavedra et al. 1997) is clearly consistent with this hypothesis, as is the observation that Tap is associated with poly(A)$^+$ RNA in living cells (Katahira et al. 1999).

If Tap is a critical mediator of mRNA export it must be recruited to mRNAs, yet recruitment of Tap to the CTE allows the nuclear export of incompletely spliced mRNAs that normally cannot exit the nucleus. These observations imply that Tap is not normally recruited to cellular mRNAs until after splicing is complete, while the CTE permits recruitment prior to splicing. In fact, there is now good evidence that mature mRNAs are bound by cellular factors that are not found on pre-mRNAs and vice versa (Luo and Reed 1999; Hir et al. 2000). Although factors selectively bound to mature mRNAs have not yet been identified in detail, they appear likely to include not only Tap but also p15. As the latter is critical for global mRNA export in yeast cells (Katahira et al. 1999), but not for CTE-dependent mRNA export in metazoan cells (Bachi et al. 2000), it appears possible that p15 may play a key role in facilitating the sequence-nonspecific recruitment of Tap to mature mRNA molecules (Fig. 4).

A possible model that incorporates these ideas is shown in Fig. 5. The initial unspliced RNA transcript is proposed to recruit a set of cellular factors that include hnRNPs and splicing commitment factors. During the process of splicing, the commitment factors, which promote nuclear retention (Legrain and

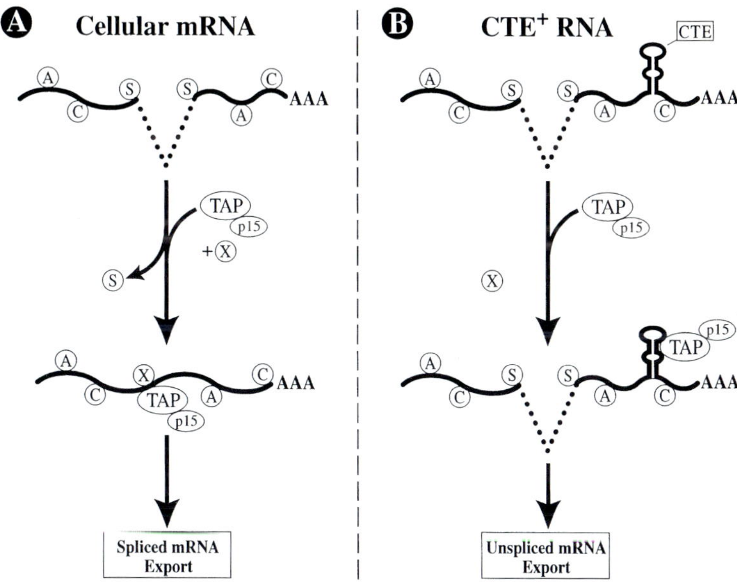

Fig. 5. Potential role of Tap in the nuclear export of mature, fully spliced mRNAs and of unspliced mRNAs bearing the CTE RNA target. See text for detailed discussion. *A or C* hnRNP proteins, *S* splicing commitment factors, *X* factors distinct from Tap and p15 that are selectively recruited to mature mRNAs

Rosbash 1989), are released while factors required for mRNA export, including Tap and p15, are coordinately recruited to the mature mRNA. The Tap protein then plays a key role in NPC targeting and cytoplasmic transport (Fig. 5).

In contrast, if the RNA contains a CTE in *cis*, Tap and most probably p15 are directly recruited prior to the release of commitment factors and probably without any requirement for the other proteins that would normally bind to mature mRNAs. This complex is then directly transported to the cytoplasm (Fig. 3) without any need for the previous release of commitment factors. In this model, the export of an unspliced vs a spliced mRNA would therefore largely depend on when Tap is recruited, thus implying that nuclear retention factors largely act by blocking export factor, and particularly Tap, recruitment.

4 Other Retroviruses, Other Pathways?

As shown in Table 1, lentiviruses and T cell leukemia viruses represent two of the three retroviral subgroups that can be classified as complex, i.e. they encode five or more gene products including a transcription factor. The third complex retrovirus family, the foamy viruses, has been less intensively studied because it does not include any pathogenic members and because human infections are very rare (reviewed by Linial 1999). Although human foamy virus (HFV), the prototype of this subgroup, does encode at least two auxiliary proteins, neither of these appears to be a Rev homologue. While it therefore seems possible that HFV contains a CTE-like element, this remains to be identified.

With the exception of type D retroviruses, very little is known about how simple retroviruses arrange for the export of their unspliced, genome-length RNA transcript. The exception to this generalization is the subgroup of avian C-type retroviruses for which the prototype is ALV (Table 1). These viruses clearly encode a CTE-like RNA export sequence that, like the MPMV CTE, is located between the *env* gene and the 3′ LTR (Fig. 1; Ogert et al. 1996; Yang and Cullen 2000). While the ALV CTE has been shown to form a complex secondary structure, the cellular target for this CTE remains unidentified. Although inhibition of Crm1 function clearly does not affect ALV CTE activity (Yang and Cullen 2000), all efforts to demonstrate a functional interaction between the ALV CTE and Tap, or a Tap/p15 complex, have thus far been unsuccessful. The future identification of the cofactor for ALV CTE-dependent RNA export therefore has the potential to shed important new light on cellular RNA transport pathways in much the same way that research into the MPMV CTE and HIV-1 Rev facilitated the identification of cellular proteins critical for mRNA and U snRNA export. Furthermore, it remains possible that other retroviral subgroups (Table 1) may utilize yet other mechanisms for the export of their intron-containing transcripts. Retroviruses will therefore clearly continue to be significant model systems for the identification and study of human nuclear RNA export pathways.

References

Amberg DA, Goldstein AL, Cole CN (1992) Isolation and characterization of *RAT1*, an essential gene of *Saccharomyces cerevisiae* required for the efficient nucleocytoplasmic trafficking of mRNA. Genes Dev 6:1173–1189

Bachi AI, Braun C, Rodrigues JP, Panté N, Ribbeck K, von Kobbe C, Kutay U, Wilm M, Görlich D, Carmo-Fonseca M, Izaurralde E (2000) The C-terminal domain of TAP interacts with the nuclear pore complex and promotes export of specific CTE-βearing RNA substrates. RNA 6:136–158

Bartel DP, Zapp ML, Green MR, Szostak JW (1991) HIV-1 Rev regulation involves recognition of non-Watson-crick base pairs in viral RNA. Cell 67:529–536

Berkhout B, Jebbink M, Zsíros J (1999) Identification of an active reverse transcriptase enzyme encoded by a human endogenous HERV-K retrovirus. J Virol 73:2365–2375

Bogerd HP, Huckaby GL, Ahmed YF, Hanly SM, Greene WC (1991) The type I human T-cell leukemia virus (HTLV-I) Rex *trans*-activator binds directly to the HTLV-I Rex and the type 1 human immunodeficiency virus Rev RNA response elements. Proc Natl Acad Sci USA 88:5704–5708

Bogerd HP, Echarri A, Ross TM, Cullen BR (1998) Inhibition of human immunodeficiency virus Rev and human T-cell leukemia virus Rex function, but not Mason-Pfizer monkey virus constitutive transport element activity, by a mutant human nucleoporin targeted to Crm1. J Virol 72:8627–8635

Braun IC, Rohrbach E, Schmitt C, Izaurralde E (1999) TAP binds to the constitutive transport element (CTE) through a novel RNA-βinding motif that is sufficient to promote CTE-dependent RNA export from the nucleus. EMBO J 18:1953–1965

Bray M, Prasad S, Dubay JW, Hunter E, Jeang K-T, Rekosh D, Hammarskjöld M-L (1994) A small element from the Mason-Pfizer monkey virus genome makes human immunodeficiency virus type 1 expression and replication Rev-independent. Proc Natl Acad Sci USA 91:1256–1260

Chang DD, Sharp PA (1989) Regulation by HIV depends upon recognition of splice sites. Cell 59:789–795

Cullen BR (1998) Retroviruses as model systems for the study of nuclear RNA export pathways. Virology 249:203–210

Ernst RK, Bray M, Rekosh D, Hammarskjöld M-L (1997) A structured retroviral RNA element that mediates nucleocytoplasmic export of intron-containing RNA. Mol Cell Biol 17:135–144

Feinberg MB, Jarrett RF, Aldovini A, Gallo RC, Wong-Staal F (1986) HTLV-III expression and production involve complex regulation at the levels of splicing and translation of viral RNA. Cell 46:807–817

Fischer U, Huber J, Boelens WC, Mattaj IW, Lührmann R (1995) The HIV-1 Rev activation domain is a nuclear export signal that accesses an export pathway used by specific cellular RNAs. Cell 82:475–483

Fornerod M, Ohno M, Yoshida M, Mattaj IW (1997a) Crm1 is an export receptor for leucine rich nuclear export signals. Cell 90:1051–1060

Fornerod M, van Deursen J, van Baal S, Reynolds A, Davis D, Murti KG, Fransen J, Grosveld G (1997b) The human homologue of yeast CRM1 is in a dynamic subcomplex with CAN/Nup214 and a novel nuclear pore component Nup88. EMBO J 16:807–816

Fridell RA, Partin KM, Carpenter S, Cullen BR (1993) Identification of the activation domain of equine infectious anemia virus Rev. J Virol 67:7317–7323

Fridell RA, Fischer U, Lührmann R, Meyer BE, Meinkoth JL, Malim MH, Cullen BR (1996) Amphibian transcription factor IIIA proteins contain a sequence element functionally equivalent to the nuclear export signal of human immunodeficiency virus type 1 Rev. Proc Natl Acad Sci USA 93:2936–2940

Fridell RA, Truant R, Thorne L, Benson RE, Cullen BR (1997) Nuclear import of hnRNP A1 is mediated by a novel cellular cofactor related to karyopherin-β. J Cell Sci 110:1325–1331

Görlich D, Kutay U (1999) Transport between the cell nucleus and the cytoplasm. Annu Rev Cell Dev Biol 15:607–660

Grassmann R, Berchtold S, Aepinus C, Ballaun C, Boehnlein E, Fleckenstein B (1991) In vitro binding of human T-cell leukemia virus Rex proteins to the Rex-response element of viral transcripts. J Virol 65:3721–3727

Grüter P, Tabernero C, von Kobbe C, Schmitt C, Saavedra C, Bachi A, Wilm M, Felber BK, Izaurralde E (1998) TAP, the human homolog of Mex67p, mediates CTE-dependent RNA export from the nucleus. Mol Cell 1:649–659

Hakata Y, Umemoto T, Matsushita S, Shida H (1998) Involvement of human CRM1 (Exportin 1) in the export and multimerization of the Rex protein of human T-cell leukemia virus type 1. J Virol 72:6602–6607

Hanly SM, Rimsky LT, Malim MH, Kim JH, Hauber J, Duc Dudon M, Le S-Y, Maizel JV, Cullen BR, Greene WC (1989) Comparative analysis of the HTLV-I Rex and HIV-1 Rev trans-regulatory proteins and their RNA response elements. Genes Dev 3:1534–1544

Heaphy S, Dingwall C, Ernberg I, Gait MJ, Green SM, Karn J, Lowe AD, Singh M, Skinner MA (1990) HIV-1 regulator of virion expression (Rev) protein binds to an RNA stem-loop structure located within the Rev response element region. Cell 60:685–693

Henderson BR, Percipalle P (1997) Interactions between HIV Rev and nuclear import and export factors: the Rev nuclear localisation signal mediates specific binding to human Importin-β. J Mol Biol 274:693–707

Hidaka M, Inoue J, Yoshida M, Seiki M (1988) Post-transcriptional regulator (Rex) of HTLV-1 initiates expression of viral structural proteins but suppresses expression of regulatory proteins. EMBO J 7:519–523

Hir HL, Moore MJ, Maquat LE (2000) Pre-mRNA splicing alters mRNP composition: evidence for stable association of proteins at exon-exon junctions. Genes Dev 14:1098–1108

Hope TJ, Bond BL, McDonald D, Klein NP, Parslow TG (1991) Effector domains of human immunodeficiency virus type 1 Rev and human T-cell leukemia virus type I Rex are functionally interchangeable and share an essential peptide motif. J Virol 65:6001–6007

Izaurralde E, Kutay U, von Kobbe C, Mattaj IW, Görlich D (1997) The asymmetric distribution of the constituents of the Ran system is essential for transport into and out of the nucleus. EMBO J 16:6535–6547

Jarmolowski A, Boelens WC, Izaurralde E, Mattaj IW (1994) Nuclear export of different classes of RNA is mediated by specific factors. J Cell Biol 124:627–635

Kadowaki T, Chen S, Hitomi M, Jacobs E, Kumagai C, Liang S, Schneiter R, Singleton D, Wisniewska J, Tartakoff AM (1994) Isolation and characterization of *Saccharomyces cerevisiae* mRNA transport-defective (*mtr*) mutants. J Cell Biol 126:649–659

Kang Y, Cullen BR (1999) The human Tap protein is a nuclear mRNA export factor that contains novel RNA-βinding and nucleocytoplasmic transport sequences. Genes Dev 13:1126–1139

Kang Y, Bogerd HP, Cullen BR (2000) Analysis of cellular factors that mediate nuclear export of RNAs bearing the Mason-Pfizer monkey virus constitutive transport element. J Virol 74:5863–5871

Katahira J, Strasser K, Podtelejnikov A, Mann M, Jung JU, Hurt E (1999) The Mex67p-mediated nuclear mRNA export pathway is conserved from yeast to human. EMBO J 18: 2593–2609

Kutay U, Bischoff FR, Kostka S, Kraft R, Görlich D (1997) Export of importin α from the nucleus is mediated by a specific nuclear transport factor. Cell 90:1061–1071

Kutay U, Lipowsky G, Izaurralde E, Bischoff FR, Schwarzmaier P, Hartmann E, Görlich D (1998) Identification of a tRNA-specific nuclear export receptor. Mol Cell 1:359–369

Legrain P, Rosbash M (1989) Some *cis*- and *trans*-acting mutants for splicing target pre-mRNA to the Cytoplasm. Cell 57:573–583

Linial ML (1999) Foamy viruses are unconventional retroviruses. J Virol 73:1747–1755

Löwer R, Tönjes RR, Korbmacher C, Kurth R, Löwer J (1995) Identification of a Rev-related protein by analysis of spliced transcripts of the human endogenous retroviruses HTDV/HERV-K. J Virol 69:141–149

Luo M-J, Reed R (1999) Splicing is required for rapid and efficient mRNA export in metazoans. Proc Natl Acad Sci USA 96:14937–14942

Magin C, Löwer R, Löwer J (1999) cORF and RcRE, the Rev/Rex and RRE/RxRE homologues of the human endogenous retrovirus family HTDV/HERV-K. J Virol 73:9496–9507

Malim MH, Cullen BR (1991) HIV-1 structural gene expression requires the binding of multiple Rev monomers to the viral RRE: Implications for HIV-1 latency. Cell 65:241–248

Malim MH, Hauber J, Fenrick R, Cullen BR (1988) Immunodeficiency virus Rev *trans*-activator modulates the expression of the viral regulatory genes. Nature 335:181–183

Malim MH, Böhnlein S, Hauber J, Cullen BR (1989a) Functional dissection of the HIV-1 Rev *trans*-activator – derivation of a *trans*-dominant repressor of Rev function. Cell 58:205–214

Malim MH, Hauber J, Le S-Y, Maizel JV, Cullen BR (1989b) The HIV-1 Rev transactivator acts through a structured target sequence to activate nuclear export of unspliced viral mRNA. Nature 338:254–257

Malim MH, McCarn DF, Tiley LS, Cullen BR (1991) Mutational definition of the human immunodeficiency virus type 1 Rev activation domain. J Virol 65:4248–4254

Mancuso VA, Hope TJ, Zhu L, Derse D, Phillips T, Parslow TG (1994) Posttranscriptional effector domains in the Rev proteins of feline immunodeficiency virus and equine infectious anemia virus. J Virol 68:1998–2001

Medstrand P, Mager DL (1998) Human-specific integrations of the HERV-K endogenous retrovirus family. J Virol 72:9782–9787

Melchior F, Paschal B, Evans E, Gerace L (1993) Inhibition of nuclear protein import by nonhydrolyzable analogs of GTP and identification of the small GTPase Ran/TC4 as an essential transport factor. J Cell Biol 123:1649–1659

Meyer BE, Malim MH (1994) The HIV-1 Rev *trans*-activator shuttles between the nucleus and the cytoplasm. Genes Dev 8:1538–1547

Moore MS, Blobel G (1993) The GTP-βinding protein Ran/TC4 is required for protein import into the nucleus. Nature 365:661–663

Neville M, Stutz F, Lee L, Davis LI, Rosbash M (1997) The importin-βeta family member Crm1p bridges the interaction between Rev and the nuclear pore complex during nuclear export. Curr Biol 7:767–775

Ogert RA, Lee LH, Beemon KL (1996) Avian retroviral RNA element promotes unspliced RNA accumulation in the cytoplasm. J Virol 70:3834–3843

Ohno M, Segref A, Bachi A, Wilm M, Mattaj IW (2000) PHAX, a mediator of U snRNA nuclear export whose activity is regulated by phosphorylation. Cell 101:187–198

Palmeri D, Malim MH (1999) Importin β can mediate the nuclear import of an arginine-rich nuclear localization signal in the absence of importin α. Mol Cell Biol 19:1218–1225

Pasquinelli AE, Ernst RK, Lund E, Grimm C, Zapp ML, Rekosh D, Hammarskjöld M-L, JE Dahlberg (1997) The constitutive transport element (CTE) of Mason-Pfizer monkey virus (MPMV) accesses a cellular mRNA export pathway. EMBO J 16:7500–7510

Patience C, Wilkinson DA, Weiss RA (1997) Our retroviral heritage. Trends Genet 13:116–120

Pollard VW, Michael WM, Naklelny S, Siomi MC, Wang F, Dreyfuss G (1996) A novel receptor-mediated nuclear protein import pathway. Cell 86:985–994

Saavedra C, Felber B, Izaurralde E (1997) The simian retrovirus-1 constitutive transport element, unlike the HIV-1 RRE, uses factors required for cellular mRNA export. Curr Biol 7:619–628

Santos-Rosa H, Moreno H, Simos G, Segref A, Fahrenkrog B, Pante N, Hurt E (1998) Nuclear mRNA export requires complex formation between Mex67p and Mtr2p at the nuclear pores. Mol Cell Biol 18:6826–6838

Segref A, Sharma K, Doye V, Hellwig A, Huber J, Lührmann R, Hurt E (1997) Mex67p, a novel factor for nuclear mRNA export, binds to both poly(a)+ RNA and nuclear pores. EMBO J 16:3256–3271

Stade K, Ford CS, Guthrie C, Weis K (1997) Exportin 1 (Crm1p) is an essential nuclear export factor. Cell 90:1041–1050

Tabernero C, Zolotukhin AS, Valentin A, Pavlakis GN, Felber BK (1996) The posttranscriptional control element of the simian retrovirus type 1 forms an extensive RNA secondary structure necessary for its function. J Virol 70:5998–6011

Tristem M (2000) Identification and characterization of novel human endogenous retrovirus families by phylogenetic screening of the human genome mapping project database. J Virol 74:3715–3730

Truant R, Cullen BR (1999) The arginine-rich domains present in human immunodeficiency virus type 1 Tat and Rev function as direct importin beta-dependent nuclear localization signals. Mol Cell Biol 19:1210–1217

Truant R, Kang Y, Cullen BR (1999) The human Tap nuclear RNA export factor contains a novel transportin-dependent nuclear localization signal that lacks nuclear export signal function. J Biol Chem 274:32167–32171

Varmus H, Brown P (1989) Retroviruses. In: Berg DE, Howe MM (eds) Mobile DNA. Am Soc Microbiol, Washington, DC, pp 53–108

Weichselbraun I, Farrington GK, Rusche JR, Böhnlein E, Hauber J (1992) Definition of the human immunodeficiency virus type 1 Rev and human T-cell leukemia virus type I Rex protein activation domain by functional exchange. J Virol 66:2583–2587

Wen W, Meinkoth JL, Tsien RY,Taylor SS (1995) Identification of a signal for rapid export of proteins from the nucleus. Cell 82:463–473

Yang J, Cullen BR (1999) Structural and functional analysis of the avian leukemia virus constitutive transport element. RNA 5:1645–1655

Yang J, Bogerd HP, Peng S, Wiegand H, Truant R, Cullen BR (1999) An ancient family of human endogenous retroviruses encodes a functional homolog of the HIV-1 Rev protein. Proc Natl Acad Sci USA 96:13404–13408

Zapp ML, Hope TJ, Parslow TG, Green MR (1991) Oligomerization and RNA binding domains of the type 1 human immunodeficiency virus Rev protein: a dual function for an arginine-rich binding motif. Proc Natl Acad Sci USA 88:7734–7738

Regulated Nuclear Transport

Christoph Schüller[1] and Helmut Ruis[1]

1 Introduction

The regulated traffic of cargoes between nucleus and cytoplasm is now thought rather to be the rule than the exception in nuclear transport. The biological importance of regulated nuclear transport is intuitively suggestive. A fast transcriptional response may be produced by rapid nuclear import of a transcription factor. Backtransfer to the cytoplasm would cease transcriptional output and the factor can be re-used. This picture is oversimplified as only few thorough studies exist. It remains to be clarified how far regulation of nuclear transport contributes to cellular signal transduction and control of its downstream targets.

This review discusses representative examples of regulated nuclear transport. We admit an overall bias towards the budding yeast *Saccharomyces cerevisiae* because this is the organism we are studying and because it is a fruitful model system for the nuclear transport field.

A number of reviews on nuclear transport have appeared recently (Mattaj and Englmeier 1998; Ohno et al. 1998; Hopper 1999; Moroianu 1999; Reiser et al. 1999a; Görner et al. 1999; Hood and Silver 1999; Kaffman and O'Shea 1999; Yamamoto and Deng 1999). Here, a short outline of general mechanisms of nuclear transport is given initially to point out potential points of regulation. One may summarize that active transport between nucleus and cytoplasm involves substrates (cargo), particularly their specific nuclear localization signals (NLS) and nuclear export signals (NES), adaptors and receptors (importins and exportins). To actively bring a substrate across the nuclear envelope, an aqueous channel is required that is part of the nuclear pore complex (NPC). Substrate, adaptors, and receptors are assembled into a transport complex before inward or outward transfer across the nuclear envelope. After dissociation of the transport complex, adaptors and receptors are recycled. Directionality of transport (import or export) will depend on the association between substrate and its receptor on one side of the nuclear envelope and dissociation on the other side. The Ran GTPase has a key role in

[1] Vienna Biocenter, Institute of Biochemistry and Molecular Cell Biology, University of Vienna, and Ludwig Boltzmann-Forschungsstelle für Biochemie, Wien, Austria

Results and Problems in Cell Differentiation, Vol. 35
K. Weis (Ed.): Nuclear Transport

generating asymmetry as discussed in much more detail elsewhere in this volume. Ran-GTP is high in the nucleus whereas Ran-GDP is predominant in the cytoplasm. This difference is thought to provide the positional information (Mattaj and Englmeier 1998).

Every step in nuclear import and export pathways could be controlled. Global control mechanisms might affect the transport capacity of nuclear pores. Changes of this kind have indeed been observed between cells in different states (Feldherr and Akin 1993) and might involve control of number or activity of nuclear pores. Global control might also involve modulation of the rate of exchange and hydrolysis of GTP. Interactions with components of the NPC could be points of global regulation. Little infor-mation is available on molecular mechanisms involved in this general control.

Specific control of nuclear concentration of individual proteins may regulate nuclear events by external signals. As a general rule, the soluble nuclear transport machinery acts fairly constitutively and is no direct target of regulation. The efficiency of transport of a protein is dependent on its interaction with components of the transport machinery. This may be controlled by masking of transport signals (NLS or NES), by complex formation, or protein modification. Modification could also have a positive effect on the cargo-transport factor interaction. Cytoplasmic or nuclear anchoring could affect transport. Anchoring involves binding of cargo molecules to a structural component, e.g., to a membrane or a cytoskeletal component by protein–protein interaction.

Many proteins imported into the nucleus will also be exported, resulting in shuttling. Altering kinetics of import, export or both will lead to a net transport of the substrate to one compartment. Control of substrate–receptor association by substrate phosphorylation is a key mechanism to alter kinetics. Interactions between a transport substrate and partners (e.g., regulatory subunits, anchoring proteins, protein kinase, or phosphatase substrates, DNA) could also be regulated by modification and influence localization. The steady-state concentration of a nuclear protein might also be influenced by controlling its stability, and this might differ in nucleus and cytoplasm. Some of these mechanisms are summarized schematically in Fig. 1.

2 Protein Kinases

Many observations have been made on the regulated nucleocytoplasmic distribution of protein kinases and the role of this distribution in cellular regulation. The nuclear transport of two groups of kinases has been studied more systematically: (1) MAP kinases, and (2) the cAMP-dependent protein kinase A. We will concentrate completely on these, and will compare the regulation of their nucleocytoplasmic localization in different classes of eukaryotes.

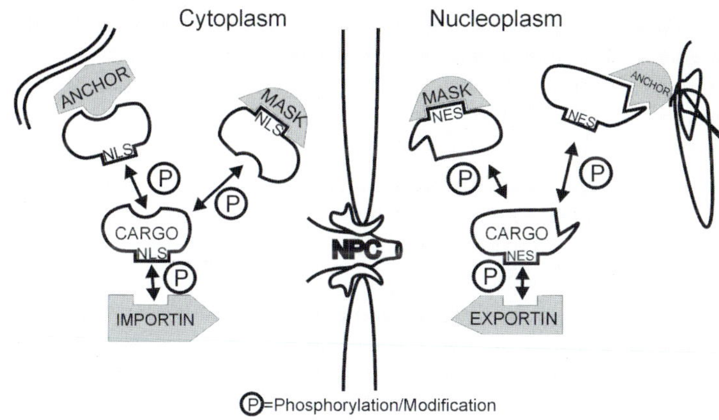

Fig. 1. Protein phosphorylation may affect various protein interactions

2.1 MAP Kinases: Control of Nuclear Concentration by Dual Phosphorylation, by Nuclear Retention and by Regulated Export

The response to a broad spectrum of extracellular stimuli is mediated by cascades of signaling proteins known as MAP kinase (MAPK) pathways. MAPK modules consist of three classes of protein kinases: the MAPK kinase kinase (MAPKKK or MEKK), which is activated by a signal sensed at the plasma membrane; this kinase phosphorylates upon activation, and activates the MAPK kinase (MAPKK or MEK), which in turn activates a MAP kinase (MAPK or ERK) by phosphorylation of a tyrosine and a threonine residue of a T–*X*–Y motif, where *X* could be glutamic acid, proline, or glycine. Control of transcription via MAPK modules usually involves MAPK-dependent modification of targets, in many cases transcription factors (Treisman 1996).

Two main classes of MAPK pathways control cell proliferation and response to stress signals. ERK pathways integrate various signals stimulating cell proliferation by growth factors (Robinson and Cobb 1997). They are also important for triggering differentiation, e.g., in neuronal cells (Cowley et al. 1994). MAPKs mediating stress responses belong to the JNK/SAPK group (Derijard et al. 1994) or to p38/RK/CSBP pathways (see e.g., Galcheva-Gargova et al. 1994; Han et al. 1994). This second class of MAPK pathways is activated by multiple environmental stresses, including osmotic stress, UV irradiation, heat stress and lipopolysaccharides, and by some cytokines (Raingeaud et al. 1995).

2.1.1 MAP Kinases Require Dual Phosphorylation for Nuclear Accumulation

Early studies on ERK1/2 (see, e.g., Lenormand et al. 1993) and on JNK1 revealed that MAPKs might be immediately involved in the transmission of regulatory signals from cytoplasm to nucleus. Details of control of this process were unravelled more recently by studies carried out to an important part on yeast MAPKs. Many features of MAPK nuclear localization appear conserved within the eukaryotic kingdom. The nuclear accumulation of ERK1/2 upon stimulation might be explained by two models: phosphorylation of MAP kinases might make them accessible for interaction with the import machinery; alternatively, active ERK1/2 might be necessary for an activating modification of (a) protein(s) participating in their nuclear accumulation. Lenormand et al. (1993) have reported that ERK phosphorylation site mutant proteins accumulate in the nucleus as efficiently as wild type proteins. However, more recently, Khokhlatchev et al. (1998) have provided evidence that ERK2 phosphorylation promotes its nuclear accumulation. These authors have also shown that phosphorylation correlates with MAPK dimerization and have suggested that dimerization may promote nuclear localization. However, whether dimerization plays a general role in nuclear accumulation remains to be addressed, and no evidence was found in the yeast systems (Gaits et al. 1998; Reiser et al. 1999b).

MAPK pathways of yeasts function in differentiation programs like the mating response, pseudohyphal development, and sporulation, in polarized growth and cell cycle progression, and in responses to stress factors (for reviews see Banuett 1998; Madhani and Fink 1998; Toone and Jones 1998). Studying yeast stress-activated MAPKs, three groups have shown that phosphorylation of the T–*X*–Y motifs of fission yeast Sty1/Spc1 MAPK and of budding yeast Hog1 MAPK is essential for stress-induced nuclear accumulation (Ferrigno et al. 1998; Gaits et al. 1998; Reiser et al. 1999b). Studies with catalytically inactive MAP kinase mutant proteins have shown that these are competent for nuclear accumulation in mammals and in yeast (Gonzalez et al. 1993; Ferrigno et al. 1998; Khokhlatchev et al. 1998; Reiser et al. 1999b). This demonstrates that MAPK-dependent modification of transport factors is not required for MAPK nuclear accumulation. Phosphorylation of MAPKs might, however, enhance nuclear import by favoring interactions with components of the import machinery.

2.1.2 MAPK Nuclear Import and Export Are Active Processes

Ferrigno et al. (1998) have shown that the budding yeast Ran homologue, Gsp1, is required for Hog1 nuclear accumulation. Gaits and Russell (1999) have shown that Sty1/Spc1 requires Pim1, a fission yeast homologue of the guanine nucleotide exchange factor RCC1. Presence of activated Sty1/Spc1 or Hog1 does not induce nuclear accumulation of mutant proteins deficient in threonine and tyrosine phosphorylation (Gaits et al. 1998; Reiser et al. 1999b), which suggests that there is no (hetero)dimer formation in these systems in connection with nuclear transport. Evidence for transport factors mediating uptake of MAPKs is lacking in most cases. Ferrigno et al. (1998) have reported that Nmd5, a budding yeast importin β homologue, is required for stress-induced Hog1 accumulation. No evidence for a preferential interaction of an import factor with phosphorylated Hog 1 has been obtained.

Nuclear export of MAPKs has been demonstrated in a number of cases (Ferrigno et al. 1998; Khokhlatchev et al. 1998; Gaits and Russell 1999; Reiser et al. 1999b). Two groups have reported that the exportin Crm1 is necessary for nuclear export of Hog1 and Sty1/Spc1 (Ferrigno et al. 1998; Gaits and Russell 1999). Evidence for physical association of a MAPK with Crm1 during export from the nucleus has been obtained for Sty1/Spc1 (Gaits and Russell 1999).

2.1.3 Nuclear Export Requires MAPK Activity and Is Correlated with MAPK Dephosphorylation

MAPK activity is required for rapid nuclear export (Ferrigno et al. 1998; Reiser et al. 1999). A catalytically inactive Hog1 K52R mutant protein is efficiently accumulated in the nucleus, but its exported is blocked. Hog1 dephosphorylation was shown to coincide with stress adaptation and nuclear export (Reiser et al. 1999b). Hog1-Y-dephosphorylation requires Ptp2 phosphatase (Wurgler-Murphy et al. 1997). Hog1-dependent activation of this phosphatase might explain the requirement of MAPK activity for export.

2.1.4 Retention Contributes to Nuclear Accumulation

If export requires MAPK dephosphorylation, stress-induced nuclear accumulation might be explained by nuclear retention of phosphorylated MAPK. Cytoplasmic anchoring and nuclear retention of MAPKs have also been suggested to control cellular MAPK localization. MEK has been proposed to act as a cytoplasmic anchor of MAPK. In such a model, the phosphorylated MAPK would be released from MEK and enter the nucleus to be retained until it is dephosphorylated. MEK has been reported to be localized preferentially in the cytoplasm because of an NES mediating its rapid nuclear export (Fukuda et al. 1997). Jaaro et al. (1997) have reported that NES deletion results in nuclear localization of MEK1 during ERK1/2 pathway stimulation. So MAPK and MEK might be co-imported into the nucleus where the MEK–MAPK complex would be disassembled after MAPK phosphorylation. MEK would then be rapidly re-exported while the MAPK would remain in the nucleus until it is dephosphorylated. Observations with yeast MAPKs would be consistent with such mechanisms. The budding yeast MEK, Pbs2, appears actively excluded from the nucleus. It has recently been demonstrated to accumulate in the nucleus when nuclear export is blocked (V. Reiser, H. Ruis, G. Ammerer, unpubl. data).

The duration of MAPK nuclear residence seems to be determined by nuclear protein interaction. Gaits et al. (1998) have observed in fission yeast that the transcription factor Atf1 is required for nuclear accumulation of Sty1/Spc1. Atf1 has been shown to be a direct target of Sty1/Spc1 (Shiozaki and Russell 1996; Wilkinson et al. 1996) and has been suggested to act as a nuclear retention factor. In budding yeast, the transcription factors Msn1, Msn2, Msn4, and Hot1 are all functionally downstream of Hog1. Their deletion has been shown to have no influence on Hog1 nuclear accumulation upon hyperosmotic stimulation, but to have a cumulative negative effect on its nuclear residence. Therefore, they act functionally like retention factors. MAPKs in higher eukaryotes appear also retained in the nucleus. Evidence for the requirement for nuclear anchor proteins not yet identified has been reported (Lenormand et al. 1998).

2.1.5 Summary

Three levels contribute to the regulation of nuclear MAPK levels: activation by phosphorylation, nuclear retention, and nuclear export requiring MAP kinase activity. However, most of the relevant molecular details of control mechanisms are still missing at all three levels. It remains to be clarified whether T–*X*–Y phosphorylation stimulates import, inhibits export, or acts in both ways. Furthermore, molecular interactions involved in nuclear retention have to be clarified. Nuclear MAPK substrates requiring phosphorylation as a precondition for nuclear export remain to be identified. The role of regulated nuclear localization in controlled signal transmission to the nucleus is plausible, but direct evidence for its impact on downstream events, particularly specific gene expression, is missing in many cases.

2.2 Protein Kinase A

The cAMP-dependent protein kinase A (PKA) is an example of a kinase whose control of localization differs dramatically from that of MAPKs. We have chosen to cover PKA localization here since it illustrates different principles and mechanisms.

2.2.1 Mammalian PKA: Extranuclear Anchoring, Diffusion to the Nucleus, and Active Nuclear Export

In its inactive form, PKA is a heterotetrameric protein consisting of two regulatory R subunits (in various RI and RII isoforms) and two catalytic C subunits (existing in three isoforms; any differences between these will be neglected in this review). On PKA pathway activation, cAMP produced by the plasma-membrane-associated adenylate cyclase binds to R subunits. This triggers dissociation of heterotetramers and formation of monomeric catalytically active C subunits. The inactive tetramer is located to various subcellular sites, e.g., the vicinity of the plasma membrane, the Golgi apparatus, mitochondria, peroxisomes, or the cytoplasm depending on the association of C subunits with RII or RI subunits and on binding to different A kinase anchoring proteins (AKAPs; for a review see Colledge and Scott 1999). Here it will only be relevant that all these sites are extranuclear. It has been postulated that, and some evidence consistent with this model has been obtained, locating the catalytically inactive enzyme to strategic sites will control preferential PKA action at predetermined sites, contributing to the specificity of intracellular events controlled by the PKA pathway. Beyond anchoring PKA, some of the AKAPs appear to act as multivalent platforms (Faux and Scott 1996) also binding protein kinase C and the protein phosphatase PP-2B. The interaction between an AKAP and the N-terminal domain of RIIa has recently been shown by solution NMR to involve an *X*-type four-helix bundle dimerization motif with an extended hydrophobic face necessary for high affinity AKAP binding (Newlon et al. 1999). Nuclear localization of AKAP95 has been reported (Eide et al. 1998), but nuclear AKAP95 appears not to interact with RIIα. AKAP95 localization was markedly changed during mitosis and appeared to overlap with that of RIIa. In interphase HeLa cells, AKAP95 has been reported to be associated with the nuclear matrix, whereas it recruits RIIa onto chromatin in an extract from mitotic cells (Collas et al. 1999). It was proposed from these studies that AKAP95 plays a role in regulating chromosome structure in mitosis. Mammalian cells disintegrate the nuclear envelope during mitosis and therefore AKAP95 does not recruit PKA to nuclei. The existence of an anchoring protein, AKAP100, which is partly nuclear, has been reported for human and rat heart. AKAP100 was co-localized with RII subunits (Yang et al. 1998a). It may therefore anchor PKA to the nucleus of some tissues.

No active transport of PKA subunits to the nucleus has been described. C subunits enter it presumably by free diffusion through nuclear pores. Translocation to the nucleus is transient, and it has been reported early that Cα subunits rapidly exit the nucleus (Nigg et al. 1985). Export is an active process mediated by binding of Cα to the PKA inhibitor protein PKI, which contains a NES (Meinkoth et al. 1993). Therefore, active PKA will appear in the cytoplasm before reaching the nucleus. In studies on memory generation it was shown that a single serotonin pulse increases the concentration of the free PKA catalytic subunit in the cytoplasm of sensory neurons. With repeated pulses, it translocates to the nucleus, where it appears to phosphorylate CREB-related transcription factors activating cAMP-inducible genes. This finding supports the hypothesis that activation of CREB-like proteins is required for the consolidation of long-term memory (Bailey et al. 1996).

2.2.2 Budding Yeast PKA: Nuclear in Rapidly Growing Cells, Nucleocytoplasmic in Slowly or Non-Growing Cells

Budding yeast PKA is encoded by three catalytic subunit genes, *TPK1–3*, and by one regulatory subunit gene, *BCY1*. Catalytic subunits are homologous to mammalian PKA C subunits. Budding yeast PKA is also a tetramer consisting of two regulatory and two catalytic subunits, dissociating upon cAMP binding to Bcy1. Bcy1 subunits are highly homologous to mammalian RII in cAMP-binding and catalytic subunit interaction, but much less so in their N-terminal domain (Toda et al. 1987) involved in AKAP interaction in the case of mammalian PKA. No proteins with significant homology to mammalian AKAPs are known in yeast. In contrast to the situation in animal cells, the budding yeast regulatory PKA subunit is entirely or partly nuclear, depending on physiological conditions. Nuclear Bcy1 localization was reported by Uno et al. (1988). It was confirmed and extended recently by Griffioen et al. (2000), who showed that PKA localization in budding yeast is under carbon source control. Both Bcy1 and the catalytic subunit Tpk1 were found concentrated in the nucleus of cells growing rapidly on glucose. Upon activation of nuclear PKA by exogenous cAMP, a rapid entry of Tpk1 into the cytoplasm was observed whereas Bcy1 remained exclusively nuclear. In contrast to the situation in rapidly growing cultures, both Bcy1 and Tpk1 are distributed over nucleus and cytoplasm in cells growing more slowly on nonfermentable carbon sources or in stationary phase cells. Griffioen et al. (2000) showed, that the Bcy1 N-terminal domain regulates Bcy1 localization. Regulation involves phosphorylation (Griffioen et al. manuscript in preparation). Phosphorylation and cytoplasmic localization was shown to depend on the Yak1 kinase, expression of which is activated by nonfermentable carbon sources or stationary conditions (Smith et al. 1998). No homologue of mammalian PKI is known in yeast. PKA relocalization to the cytoplasm requires Zds1, a protein with many functions (Bi and Pringle 1996; Yu et al. 1996), which appears to interact with Bcy1. Zds1 might act as a functional AKAP analog.

3 Transcription Factors

Many transcription factors appear regulated at the level of nuclear transport. Here we will present selected examples of systems that have been analyzed in some detail.

3.1 Swi5: Regulated Nuclear Import and Nuclear Degradation

Cell division in budding yeast is asymmetric resulting in a larger mother and a smaller daughter cell. In contrast to daughter cells, about 70% of the mother cells change their mating type upon cell division of homothallic yeast. This event includes expression of an endonuclease (HO) and the subsequent rearrangement of the mating-type locus. The transcription factor Swi5 is important for mating-type switching as it controls expression of HO. Expression of Swi5 is restricted to S, G2, and M phases. At the end of mitosis, Swi5 enters the nucleus, activates the HO gene and is rapidly degraded (Nasmyth et al. 1990). The Swi5 NLS was sufficient to confer cell cycle dependent localization onto a β-galactosidase reporter protein. Nuclear Swi5 translocation is controlled by the cyclin/CDK Cdc28, whose activity declines at the exit of mitosis. Two of the three Cdc28 phosphorylation sites of Swi5 are located within its NLS, the third is close to it. Mutation of phosphorylation site serine residues to alanines renders the factor progressively more nuclear. The drop of CDK activity of Cdc28 allows the Cdc14 phosphatase to activate the NLS by dephosphorylation (Moll et al. 1991). In the nucleus, Swi5 binds to the HO promoter in mother and daughter cells, but is removed in daughter cells by the repressor Ash1 before the endonuclease is expressed. Ash1 is expressed only in daughter cells because of asymmetric localization of its mRNA. In mother cells, which lack nuclear Ash1, Swi5 is degraded simultaneously with activation of the HO promoter (Cosma et al. 1999). The concerted action of an NLS activated in a cell cycle-dependent manner by dephosphorylation, interaction with a repressor involved in mother–daughter control and nuclear degradation causing irreversible transcription factor inactivation yields a sharp peak of nuclear activity.

3.2 Pho4: Switch-Like Regulation of a Shuttling Protein by Phosphorylation

Budding yeast Pho4 is one of the best understood examples of a transcription factor whose nuclear localization is regulated. It is the key regulator of the response to limiting amounts of inorganic phosphate. When phosphate is present in the growth medium, Pho4 is kept phosphorylated by a protein kinase of the cyclin/CDK type (Pho80/Pho85) and is mainly cytoplasmic. Upon phosphate starvation, the Pho80/Pho85 CDK is inhibited by an inhibitory subunit (CDI), Pho81. Consequently, Pho4 is dephosphorylated by an unidentified protein phosphatase (Kaffman et al. 1994). These events coincide with Pho4 nuclear accumulation, its binding to promoters, and activation of specific transcription. In the simplest model, Pho4 phosphorylation would determine its intracellular localization and transcription factor activity. However, a thorough analysis of the localization behavior of Pho4 in a set of karyopherin mutants revealed a more complex situation.

Budding yeast cells lacking the exportin Msn5 are viable. Msn5 has features typical for an importin β homologue (Görlich et al. 1997). Mutants defective in Msn5 show a variety of phenotypes reflecting its function in several different signal transduction pathways (Alepuz et al. 1999). *msn5* mutant cells show a constitutive nuclear localization of Pho4. Msn5 is therefore necessary for Pho4 nuclear export in vivo. Further analysis showed that it binds selectively to phosphorylated Pho4 in vitro. According to this, Msn5 functions as the Pho4 exportin and mediates its export when phosphate is abundant and the Pho80/Pho85 kinase is active (Kaffman et al. 1998a). The importin β homologue Pse1/Kap121 is necessary for Pho4 import (Kaffman et al. 1998b). In this case the biochemical analysis showed that Pho4 phosphorylation interferes with cargo recognition and that phosphorylation of a particular serine in a basic stretch of amino acid residues interferes with binding of Pse1/Kap121 to Pho4 in vitro. Taken together, phosphorylation of Pho4 stimulates nuclear export and inhibits import.

However, a third level of regulation controls transcriptional activity of nuclear Pho4. The identification of the Pho4 exportin allowed determination of the transcriptional output triggered by trapping it constitutively in the nucleus. Surprisingly, no effect on the target gene *PHO5* was detectable. Again the phosphorylation status of Pho4 was shown to be crucial for regulation. Pho2, a factor related to Pho4 and required for efficient transcription of several genes of the Pho regulon, associates with Pho4 in a phosphorylation-dependent manner. Only the heterodimer is a highly active transcription factor (Komeili and O'Shea 1999). In summary, Pho4 is a transcription factor regulated by its phosphorylation status, which influences its interaction with other proteins. Regulation of import, export and DNA-binding are mediated by separable domains. Dual regulation of import and export creates a molecular switch.

3.3 Mig1 and Msn2/4: Variations of the Pho4 Theme?

As illustrated above, modification sites controlling nuclear export can be broadly scattered over the cargo and may cooperate in an additive fashion. The same kinase may regulate several aspects of transcription factor function, i.e., its nuclear import, export or DNA binding. Two more yeast transcription factors seem regulated similarly.

The C_2H_2 Zn-finger protein Mig1 functions as a transcriptional repressor involved in regulation of many genes in response to glucose in the medium. Under these conditions, the AMP-activated protein kinase Snf1 is inactive and Mig1, a Snf1 substrate (Treitel et al. 1998), is localized in the nucleus. Upon carbon source starvation in glucose-grown cells Mig1 is rapidly phosphorylated and exported (DeVit et al. 1997). Serine residues that are target sites for Snf1 are important for regulated transport. Mutation of these revealed an almost linearly cumulative phenotype with complete loss of export upon replacing four serines by alanine. The exportin Msn5 is necessary for Mig1 nuclear export. Identification of the export factor revealed a mechanism of export control similar to that of Pho4 (DeVit et al. 1999). No information on the mechanism of Mig1 import and therefore on its regulation is available.

The functionally redundant *Saccharomyces cerevisiae* transcription factors Msn2 and Msn4 activate transcription in response to stress (Martinez-Pastor et al. 1996) They provide an example for a dynamically regulated, rapid, and reversible translocation response. Both change their localization in response to stress and PKA activity (Görner et al. 1998). Although Msn2 and Msn4 are mainly cytoplasmic in unstressed cells, both factors accumulate in the nucleus within minutes upon exposure to stress. In cells with low PKA activity, the factors also become nuclear, whereas high PKA activity triggers rapid export even under sustained stress conditions (Görner et al. 1998). This regulatory system seems to be based on a balance between PKA and stress factors. PKA activity influences both strength of export and import domains. High PKA leads to inactivation of the import domain, presumably by masking it by phosphorylation of a cluster of four PKA sites, whereas it allows interaction of the export domain with Msn5. Because studies in *msn5* mutants have demonstrated that constitutively nuclear Msn2 is still regulated by stress and PKA (Schüller et al., in prep.), Msn2, like Pho4, is regulated at the levels of nuclear import, export and at an intranuclear level.

3.4 NF-κB: Control by NLS Masking and Piggyback Nuclear Export

The mammalian transcription factor NF-κB is held in the cytoplasm by its retention factor IκB. Upon a broad variety of stimuli, IκB is phosphorylated and degraded, allowing NF-κB to accumulate in the nucleus (Siebenlist et al. 1995). NF-κB is controlled by a negative feedback mechanism leading to nuclear export. It activates the transcription of IκB, which is able to shut off NF-κB-dependent transcription by entering the nuclear compartment, displacing NF-κB from DNA and exporting it to the cytoplasm (Arenzana-Seisdedos 1995). Nuclear import of IκB not bound to NF-κB involves importin (α) and (β) and cytosolic factors interacting with its ankyrin repeats (Turpin et al. 1999). The leucine-rich NES of IκB is thought to mediate the export of a nuclear NF-κB/IκB complex. IκB blocks NF-κB import by masking its NLS. Recently the structure of a NF-κB/IκB complex was reported (Huxford at al. 1998; Jacobs and Harrison 1998) and substantially supported this model by revealing a close contact between IκB and the NF-κB NLS. Results obtained with this system have clarified details of protein interaction with the function of NLS-masking, furthermore that protein phosphorylation may also act at the level of a masking protein, and that nuclear export may be mediated by an interaction partner providing an NES.

3.5 p53: NES Masking by Homotetramerization Versus Synergistic Effects on Export by Heterodimer Formation?

The p53-mediated stress response is frequently impaired in human cancer. Cytoplasmic sequestration of p53 probably contributes to tumorigenesis. p53 is nuclear in G1 and cytosolic in G2 and S phase. In response to stress it moves into the nucleus, gets stabilized and induces cell cycle arrest or apoptotic cell death depending on the amount of DNA damage and on cell type. In unstressed cells, the oncoprotein Mdm2 binds to p53 and facilitates cytoplasmic degradation (Momand et al. 2000). Mdm2 is a shuttling protein. It possesses a leucine-rich NES and might contribute to p53 nuclear export by binding p53 in the nucleus and exporting it in a complex. It has been determined that NES sequences in p53 and Mdm2 are relatively weak (Henderson and Eleftheriou 2000). Complex formation might enhance their overall efficiency. How is p53 retained in the nucleus in stressed cells? Phosphorylation regulates its DNA-binding activity by promoting tetramer formation (Lakin and Jackson 1999). Structural arguments and the overlap of the p53 NES with the tetramerization domain point to a mechanism where intranuclear interaction occludes the NES in the tetramer, the DNA-binding form of p53 (Stommel et al. 1999). Therefore, stress-induced modification enhances DNA-binding affinity, and inhibits nuclear export and cytoplasmic degradation. It is interesting to note that Mdm2 itself is upregulated by p53, which creates an autoregulated feedback loop similar to the situation found with NF-κB. This example illustrates that NES elements can be enhanced by heterodimer formation while they may be masked by homo-oligomer interactions.

3.6 NF-AT: NLS Masking by Phosphorylation and Nuclear Export Control by Exportin–Phosphatase Competition

The NF-AT family comprises four related transcription factors sharing a C-terminal Rel homology domain. N-terminally, these factors share a calcineurin binding domain and sequences responsible for transport regulation. Calcium influences gene expression in many different cells (Clapham 1995). The Ca^{2+}-calcineurin pathway stimulates rapid nuclear translocation of NF-AT in lymphoid cells, in turn activating immune-response genes. Both nuclear import and export of NF-AT seem to be regulated by phosphorylation. Nuclear import of NF-AT2 was shown to be counteracted by GSK3 in a mechanism involving the intramolecular masking of two NLS sequences, one near the N-terminus, the other at the C-terminus (Beals et al. 1997a,b). The other NF-ATs have a slightly different organization of their regulatory transport domain. NF-AT4 export appears regulated by a mechanism involving direct competition of calcineurin with the exportin Crm1 (Zhu and McKeon 1999). Calcineurin binds in a noncatalytic manner to the region also containing a canonical NES. It remains to be seen if this is also true for other members of the NF-AT family. NF-AT studies show that phosphorylation may also regulate intramolecular masking of NLS and that transport domain-masking may play a role at the level of nuclear export.

The *S. cerevisiae* NF-AT ortholog Crz1/Tcn1 also acts downstream of calcineurin. Elevated calcium levels cause its rapid but transient nuclear import. This effect is linked to dephosphorylation of Crz1/Tcn1 by calcineurin (Stathopoulos-Gerontides et al. 1999).

3.7 Yap1: Modification by Oxidation as an Alternative to Phosphorylation?

Budding yeast Yap1 is a member of the bZIP-family. It is involved in the regulation of a relatively large regulon in response to oxidative stress (Lee et al. 1999). Upon treatment of cells with hydrogen peroxide, a Yap1–GFP fusion protein accumulates in the nucleus within minutes. The import of Yap1 seems to be constitutive, but is an active process dependent on the Ran-system (Yan et al. 1998). A cysteine rich domain (CRD) at the Yap1 C-terminus was identified and shown to be sufficient to confer oxidative stress-regulated localization (Kuge et al. 1997). Later studies indicated that Yap1 localization might involve regulated nuclear export (Kuge et al. 1998; Yan et al. 1998). The exportin Crm1 can be immunoprecipitated with Yap1 in a complex with Gsp1–GTP (the yeast Ran homologue) but not with Gsp1–GDP. Formation of this complex is significantly favored in the presence of reducing agent. In a two hybrid assay, Crm1 interacts with Yap1 or its CRD. This interaction is sensitive to oxidation. Three cysteine residues within or near the CRD are important for its function and regulation. Taken together, the data suggest a model whereby oxidation of cysteines disrupts Crm1 interaction causing cessation of export. The Yap1 CRD is only remotely related to the canonical NES. This might be necessary to provide weak interaction with Crm1, which might favor regulation by modification (Yan et al. 1998). The apparent direct redox regulation of Yap1 export is an outstanding exception from the common theme of regulation by phosphorylation, which, however, has not been rigorously excluded in the Yap1 system.

3.8 HIF1: Hypoxia Control of an NLS and of an Overlapping Activation Domain

Transcription of many genes is regulated by oxygen availability and influences profoundly processes such as angiogenesis, erythropoiesis, and glycolysis. The hypoxia-inducible factor HIF1 α (Kallio et al. 1998) is regulated by nuclear import in response to hypoxic conditions, its interaction with ARNT and by stabilization of the protein. This factor carries two NLS motifs, one of them located near the N-terminus within the DNA binding domain and acting constitutively. The bHLH domain of this and other factors (also ARNT and the dioxin receptor; Ikuta et al. 1998) confers constitutive nuclear localization. At the C-terminus two regulatory domains overlap, a second NLS and an activation domain that recruits the coactivator CBP/p300. Both of these two latter elements are regulated by oxygen levels. Nuclear localization of the factor per se is not sufficient for induction of transcription, as a constitutive nuclear version still showed hypoxia-induced transcription and CBP recruitment. Therefore, HIF1 α is regulated on at least three levels: stabilization, localization, and activation of transcription (Kallio et al. 1998). This provides another example showing that nuclear localization may not be sufficient for activation of transcription and shows that NLS elements may overlap with other functional domains, in this case with an activation domain.

3.9 Cyclin B1: Piggyback Nuclear Uptake and an NES Overlapping with a Cytoplasmic Retention Signal

M-phase-promoting factor (MPF), a complex of cdc2 and a B-type cyclin, regulates the G2/M cell cycle transition. Cyclin B1 accumulates in the cytoplasm through S and G2 phases and translocates to the nucleus during prophase. Nuclear exclusion of cyclin B1 is important in the control of the DNA damage-induced G2 checkpoint. Regulation of B1 localization was postulated to include a cytoplasmic retention mechanism (Pines and Hunter 1994). Also here, phosphorylation was found to regulate intracellular localization. Later studies defined a nuclear export signal overlapping with the domain containing the cytoplasmic retention activity (Yang et al. 1998b). In fact, cytoplasmic localization of cyclin B1 is regulated at the nuclear export level by a Crm1-dependent mechanism. Therefore, cytoplasmic localization is the effect of a regulated nuclear export domain. Although no consensus NLS could be defined in both MPF subunits, it was shown by two hybrid analysis that Cyclin F is probably responsible for transport. It contains two nuclear targeting signals and binds to B1, thereby contributing the nuclear import signal to the complex (Kong et al. 2000).

3.10 Wnt Signaling to the Nucleus by Preventing β-Catenin Degradation

Wnts are secreted glycoproteins that act as ligands to stimulate receptor-mediated signal transduction pathways in both vertebrates and invertebrates (Moon et al. 1997). The Wnt/β-catenin pathway is the best understood Wnt signaling pathway, and is highly conserved during evolution. In the absence of Wnt signaling, APC ("adenomatous polyposis coli"), β-catenin, and axin are phosphorylated by GSK-3. All these components are constituents of a cytoplasmic degradation complex. It promotes interaction of β-catenin with beta-TrCP, leading to the ubiquitination of β-catenin and its degradation by the proteasome. Thus, in the absence of Wnt signaling, β-catenin is rapidly degraded in the cytoplasm (Latres et al. 1999). The active Wnt pathway seems to decrease the activity of GSK-3β and APC, thereby increasing the level of β-catenin. This functions then as an adapter linking cadherins (at the plasma membrane) and, after translocation to the nucleus, as a coactivator for transcription with LEF1/TCF (Eastman and Grosschedl 1999).

3.11 Notch: Release of a Soluble Transcription Factor by Proteolytic Cleavage of a Transmembrane Protein

Notch proteins are large (300 kDa) single span transmembrane proteins involved in many developmental decisions (Artavanis-Tsakonas et al. 1999). They are activated by binding of transmembrane ligands such as Delta (from *Drosophila*). This system is somewhat different from other signaling cascades, as binding of the ligand leads to proteolytic cleavage of Notch in the transmembrane domain, which produces a cytoplasmic fragment (Schroeter et al. 1998). This intracellular domain (ICD) acts as transcriptional activator and directly regulates transcription after translocation to the nucleus and association with DNA-binding proteins such as SuH ("suppressor of hairless"). Complexes of ICD with SuH are found to reside to some degree in the cytoplasm despite ICD having a functional NLS (Kidd et al. 1998). Therefore, it was proposed that the access of ICD to the nucleus is also regulated, allowing signal output only after a certain threshold is achieved.

3.12 Light Regulation in Plants: Light-Induced Nuclear Localization of a Photoreceptor and Light Activated Nuclear Export of a Repressor

Light exerts profound influence on plant development (Yamamoto and Deng 1999). Seedlings grown in the light follow a pathway of photomorphogenesis opposed to skotomorphogenesis in the dark. Light signals are detected by photoreceptors, e.g., by phytochromes A to E. Regulated nuclear localization plays an important role in light signaling. The photoreceptor phytochrome B (PhyB) is localized in the nucleus in the light but absent from it in the dark (e.g., Kircher et al. 1999). Recently, it has been shown in a two-hybrid assay using the PhyB C-terminus as bait that PhyB binds to the transcription factor PIF3. Only the light-activated form of PhyB interacts with PIF3 in vitro. The interaction between PIF3 and PhyB is also light sensitive. PIF3, a nuclear factor, in turn activates other transcription factors possibly involved in the plant circadian clock (Martinez-Garcia et al. 2000). Therefore, phytochromes might act as simple photoswitchable elements. The mirror example to light induced nuclear import is the photomorphogenic regulator COP1, which accumulates in the nucleus in dark-grown cells and acts as repressor (von Arnim and Deng 1994). Light causes COP1 export involving a regulated NES (Stacey et al. 1999). COP1 might also be retained in the cytoplasm by the interacting factor CIP1 which associates with the cytoskeleton and may act as a cytoplasmic anchor (Matsui et al. 1995).

3.13 Glucocorticoid Receptor (GR): Control of Nuclear Import by Dissociation of a Complex with HSP – or Shuttling of a GR–HSP Complex?

In the absence of hormone, steroid receptors are packaged into HSP90-containing 8S complexes. Binding of hormone dissociates these and results in specific binding of steroid hormone receptors to DNA and transcriptional activation. Some unliganded steroid receptors are constitutively nuclear (e.g., progesterone receptor, PR) whereas others are predominantly cytoplasmic (GR) and undergo nuclear import upon ligand binding. This simple pattern is incomplete as steroid hormone receptors shuttle continuously (e.g., PR; Guiochon-Mantel et al.. 1991) Unexpectedly, also the HSP90-containing 8S complexes localized in the nucleus are able to leave it rapidly according to heterokaryon assays. Further, it was shown for GR that, upon ligand withdrawal, the 8S complex is formed in the nucleus and is slowly exported (Haché et al. 1999, and references therein). The 8S complex could also be transferred artificially to the nucleus by the addition of a nuclear retention signal. Therefore, the localization of steroid receptors is probably determined by relative nuclear import and export rates and seems independent of associated HSPs.

4 Stress Regulation of Nuclear mRNA Export

The basic mechanism of mRNA nuclear export is discussed elsewhere in this volume and has been reviewed (Sträßer and Hurt 1999). Different RNA species follow specific export routes. Yeast cells exposed to a severe stress (42°C or 10% ethanol) shut down synthesis of most proteins with the exception of stress proteins. This effect is partly caused by a general block of poly(A) mRNA export, which is bypassed by heat shock mRNAs (Saavedra et al. 1996). Export of heat shock mRNAs does not depend on Ran, but rather involves other export factors including Rip1 and Gle1 (Saavedra et al. 1997). The shut-off of mRNA export by stress involves the dissociation of Npl3, a key mediator of poly(A) mRNA export, from mRNA under these conditions (Krebber et al. 1999).

5 Some General Conclusions

Many eukaryotic proteins show differential nucleocytoplasmic localization. Regulatory mechanisms involved include the interaction with karyopherins (importins and exportins), with masking factors, anchoring proteins, with DNA-binding proteins and presumably with DNA itself. Control of these interactions is exerted by protein phosphorylation in most cases, usually on serine or threonine residues of cargo proteins. Many nuclear proteins are shuttling between cytoplasm and nucleus. In some cases, it has been demonstrated that proteins, whose local steady state concentration in nuclei and cytoplasm does not change, are nevertheless shuttling continuously between these two compartments. Any changes in local concentration to be achieved by regulation may then be triggered by affecting in various ways the kinetics of import, export or both.

What does a conditional localization tell us? Observation of a change in localization indicates ongoing signaling events. Intracellular distribution of a protein is usually assayed by either in situ indirect immunofluorescence or more directly by direct observation of green fluorescent protein (GFP) fusion proteins, whereas transport is usually detected by following GFP fusions, by heterokaryon assays, or by microinjection. When interpreting such assays, one should bear in mind that the methods of detection are relatively insensitive and that few molecules of a transcription factor or another signaling molecule may be sufficient to produce a significant transcriptional output. Therefore, to make direct conclusions from localization studies on overall output (e.g., gene expression) is tempting, but not justified.

The fact that transport regulation may not be tight enough in some cases may be one reason why organisms have evolved multiple levels of regulation: together, these allow more stringent overall control. Generally, multistep processes with several points of control will also permit integration of multiple input signals and to fine tune the response to them to create a proper and dynamic output.

References

Alepuz PM, Matheos D, Cunningham KW, Estruch F (1999) The *Saccharomyces cerevisiae* RanGTP-binding protein msn5p is involved in different signal transduction pathways. Genetics 153:1219–1231

Arenzana-Seisdedos F, Thompson J, Rodriguez MS, Bachelerie F, Thomas D, Hay RT (1995) Inducible nuclear expression of newly synthesized I kappa B alpha negatively regulates DNA-binding and transcriptional activities of NF-kappa B. Mol Cell Biol 15:2689–2696

Artavanis-Tsakonas S, Rand MD, Lake RJ (1999) Notch signaling: cell fate control and signal integration in development. Science 284:770–776.

Bailey CH, Bartsch D, Kandel RR (1996) Toward a molecular definition of long-term memory storage. Proc Natl Acad Sci USA 93:13445–13452

Banuett F (1998) Signalling in the yeasts: an informational cascade with links to the filamentous fungi. Microbiol Mol Biol Rev 62:249–274

Beals CR, Clipstone NA, Ho SN, Crabtree GR (1997a) Nuclear localization of NF-ATc by a calcineurin-dependent, cyclosporin-sensitive intramolecular interaction. Genes Dev 11:824–834.

Beals CR, Sheridan CM, Turck CW, Gardner P, Crabtree GR (1997b) Nuclear export of NF-ATc enhanced by glycogen synthase kinase-3. Science 275:1930–1934

Bi E, Pringle JR (1996) *ZDS1* and *ZDS2*, genes whose products may regulate Cdc42p in *Saccharomyces cerevisiae*. Mol Cell Biol 16:5264–5275

Clapham DE (1995) Calcium signaling. Cell 80:259–268

Collas P, Le Guellec K, Tasken K (1999) The A-kinase-anchoring protein AKAP95 is a multivalent protein with a key role in chromatin condensation at mitosis. J Cell Biol 147:1167–1180

Colledge M, Scott JD (1999) AKAPs: from structure to function. Trends Cell Biol 9:216–221

Cosma MP, Tanaka T, Nasmyth K (1999) Ordered recruitment of transcription and chromatin remodeling factors to a cell cycle- and developmentally regulated promoter. Cell 97:299–311

Cowley S, Paterson H, Kemp P, Marshall CJ (1994) Activation of MAP kinase is necessary and sufficient for PC12 differentiation and for transformation of NIH 3T3 cells. Cell 77:841–852

Derijard B, Hibi M, Wu IH, Barrett T, Su B, Deng T, Karin M, Davis RJ (1994) JNK1: a protein kinase stimulated by UV light and Ha-Ras that binds and phosphorylates the c-Jun activation domain. Cell 76:1025–1037

DeVit MJ, Johnston M (1999) The nuclear exportin Msn5 is required for nuclear export of the Migl glucose repressor of *Saccharomyces cerevisiae*. Curr Biol 9:1231–1241

DeVit MJ, Waddle JA, Johnston M (1997) Regulated nuclear translocation of the Migl glucose repressor. Mol Biol Cell 8:1603–1618.

Eastman Q, Grosschedl R (1999) Regulation of LEF-1/TCF transcription factors by Wnt and other signals. Curr Opin Cell Biol 11:233–240

Eide T, Coghlan V, Orstavik S, Holsve C, Solverg R, Skalhegg BS, Lamb NJ, Langeberg L (1998) Molecular cloning, chromosomal localization, and cell cycle-dependent subcellular distribution of the A-kinase anchoring protein, AKAP95. Exp Cell Res 238:305–316

Faux MC, Scott JD (1996) Molecular glue: kinase anchoring and scaffold proteins. Cell 85:9–12

Feldherr CM, Akin D (1993) Regulation of nuclear transport in proliferating and quiescent cells. Exp Cell Res 205:179–186

Ferrigno P, Posas F, Koepp D, Saito H, Silver PA (1998) Regulated nucleo/cytoplasmic exchange of HOGl MAPK requires the importin beta homologs NMD5 and XPO1. EMBO J 17: 5606–5614

Freedman DA, Levine AJ (1998) Nuclear export is required for degradation of endogenous p53 by MDM 2 and human papillomavirus E6. Mol Cell Biol 18:7288–7293

Fukuda M, Gotoh I, Adachi M, Gotoh Y, Nishida E (1997) A novel regulatory mechanism in the mitogen-activated protein (MAP) kinase cascade. Role of nuclear export signal of MAP kinase. J Biol Chem 272:32642–32648

Gaits F, Russell P (1999) Active nucleocytoplasmic shuttling required for function and regulation of stress-activated kinase Spcl/Styl in fission yeast. Mol Biol Cell 10:1395–1407

Gaits F, Degols G, Shiozaki K, Russell P (1998) Phosphorylation and association with the transcription factor Atfl regulate localization of Spcl/Styl stress-activated kinase in fission yeast. Genes Dev 12:1464–1473

Galcheva-Gargova Z, Derijard B, Wu IH, Davis RJ (1994) An osmosensing signal transduction pathway in mammalian cells. Science 265:806–808

Gonzalez FA, Seth A, Raden DL, Bowman DS, Fay FS, Davis RJ (1993) Serum-induced translocation of mitogen-activated protein kinase to the cell surface ruffling membrane and the nucleus. J Cell Biol 122:1089–1101

Görlich D, Dabrowski M, Bischoff FR, Kutay U, Bork P, Hartmann E, Prehn S, Izaurralde E (1997) A novel class of RanGTP binding proteins. J Cell Biol 138:65–80

Görner W, Durchschlag E, Martinez-Pastor MT, Estruch F, Ammerer G, Hamilton B, Ruis H, Schuller C (1998) Nuclear localization of the C2H2 zinc finger protein Msn2p is regulated by stress and protein kinase A activity. Genes Dev 12:586–97.

Görner W, Schüller C, Ruis H (1999) Being at the right place at the right time: the role of nuclear transport in dynamic transcriptional regulation in yeast. Biol Chem 380:147–150

Griffioen G, Anghileri P, Imre E, Baroni MD, Ruis H (2000) Nutritional control of nucleocytoplasmic localization of cAMP-dependent protein kinase catalytic and regulatory subunits in *Saccharomyces cerevisiae*. J Biol Chem 275:1449–1456

Guiochon-Mantel A, Lescop P, Christin-Maitre S, Loosfelt H, Perrot-Applanat M, Milgrom E (1991) Nucleocytoplasmic shuttling of the progesterone receptor. EMBO J 10:3851–3859

Hache RJ, Tse R, Reich T, Savory JG, Lefebvre YA (1999) Nucleocytoplasmic trafficking of steroid-free glucocorticoid receptor J Biol Chem 274:1432–1439

Han J, Lee JD, Bibbs L, Ulevitch RJ (1994) A MAP kinase targeted by endotoxin and hyperosmolarity in mammalian cells. Science 265:808–811

Henderson BR, Eleftheriou A (2000) A comparison of the activity, sequence specificity, and CRM1-dependence of different nuclear export signals. Exp Cell Res 256:213–24

Hopper AK (1999) Nucleocytoplasmic transport: inside out regulation. Curr Biol 9:R803–R806

Hood JK, Silver PA (1999) In or out? Regulating nuclear transport. Curr Opin Cell Biol 11:241–247

Huxford T, Huang DB, Malek S, Ghosh G (1998) The crystal structure of the IkappaBalpha/NF-kappaB complex reveals mechanisms of NF-kappaB inactivation. Cell 95:759–770

Ikuta T, Eguchi H, Tachibana T, Yoneda Y, Kawajiri K (1998) Nuclear localization and export signals of the human aryl hydrocarbon receptor. J Biol Chem 273:2895–2904

Jaaro H, Rubinfeld H, Hanoch T, Seger R (1997) Nuclear translocation of a mitogen-activated protein kinase (MEK1) in response to mitogenic stimulation. Proc Natl Acad Sci USA 94: 3742–3747

Jacobs MD, Harrison SC (1998) Structure of an IkappaBalpha/NF-kappaB complex. Cell 95: 749–758

Kaffman A, O'Shea EK (1999) Regulation of nuclear localization: a key to a door. Annu Rev Cell Dev Biol 15:291–339

Kaffman A, Herskowitz I, Tjian R, O'Shea EK (1994) Phosphorylation of the transcription factor PHO4 by a cyclin-CDK complex, PHO80-PHO85. Science 263:1153–1156

Kaffman A, Rank NM, O'Shea EK (1998a) Phosphorylation regulates association of the transcription factor Pho4 with its import receptor Pse1/Kap121. Genes Dev 12:2673–2683

Kaffman A, Rank NM, O'Neill, EM, Huang LS, O'Shea EK (1998b) The receptor Msn5 exports the phosphorylated transcription factor Pho4 out of the nucleus. Nature 396:482–486

Kallio PJ, Okamoto K, O'Brien S, Carrero P, Makino Y, Tanaka H, Poellinger L (1998) Signal transduction in hypoxic cells: inducible nuclear translocation and recruitment of the CBP/p300 coactivator by the hypoxia-inducible factor-1alpha. EMBO J 17:6573–6586

Kidd S, Lieber T, Young MW (1998) Ligand-induced cleavage and regulation of nuclear entry of Notch in *Drosophila melanogaster* embryos. Genes Dev 12:3728–3740

Kircher S, Kozma-Bognar L, Kim L, Adam E, Harter K, Schafer E, Nagy F (1999) Light quality-dependent nuclear import of the plant photoreceptors phytochrome A and B. Plant Cell 11: 1445–1456

Kokhlatchev AV, Canagarajah B, Wilsbacher J, Robinson M, Atkinson M, Goldsmith E, Cobb MH (1998) Phosphorylation of the MAP kinase ERK2 promotes its homodimerization and nuclear translocation. Cell 93:605–615

Komeili A, O'Shea EK (1999) Roles of phosphorylation sites in regulating activity of the transcription factor Pho4. Science 284:977–980

Kong M, Barnes EA, Ollendorff V, Donoghue DJ (2000) Cyclin F regulates the nuclear localization of cyclin B1 through a cyclin–cyclin interaction. EMBO J 19:1378–1388

Krebber H, Taura T, Lee MS, Silver PA (1999) Uncoupling of the hnRNP Npl3p from mRNAs during the stress-induced block in mRNA export. Genes Dev 13:1994–2004

Kuge S, Jones N, Nomoto A. (1997) Regulation of yAP-1 nuclear localization in response to oxidative stress. EMBO J 16:1710–1720

Kuge S, Toda T, Iizuka N, Nomoto A (1998) Crm1 (XpoI) dependent nuclear export of the budding yeast transcription factor yAP-1 is sensitive to oxidative stress. Genes Cells 3:521–532

Lakin ND, Jackson SP (1999) Regulation of p53 in response to DNA damage. Oncogene 18: 7644–7655

Latres E, Chiaur DS, Pagano M (1999) The human F box protein beta-Trcp associates with the Cul1/Skp1 complex and regulates the stability of beta-catenin. Oncogene 18:849–854

Lee J, Godon C, Lagniel G, Spector D, Garin J, Labarre J, Toledano MB (1999) Yap1 and Skn7 control two specialized oxidative stress response regulons in yeast. J Biol Chem 274:16040–16046

Lenormand P, Sardet C, Pages G, L'Allemain G, Brunet A, Pouyssegur J (1993) Growth factors induce nuclear translocation of MAP kinases (p42mapk and p44mapk) but not of their activator MAP kinase (p45mapkk) in fibroblasts. J Cell Biol 122:1079–1088

Lenormand P, Brondello JM, Brunet A, Pouyssegur J (1998) Growth factor-induced p42/p44 MAPK nuclear translocation and retention requires both MAPK activation and neosynthesis of nuclear anchoring proteins. J Cell Biol 142:625–633

Madhani HD, Fink GR (1998) The riddle of MAP kinase signaling specificity. Trends Genet 14: 151–155

Martinez-Garcia JF, Huq E, Quail PH (2000) Direct targeting of light signals to a promoter element-bound transcription factor. Science 288:859–863

Martinez-Pastor MT, Marchler G, Schüller C, Marchler-Bauer A, Ruis H, Estruch F (1996) The *Saccharomyces cerevisiae* zinc finger proteins Msn2p and Msn4p are required for transcriptional induction through the stress response element (STRE). EMBO J 15:2227–2235

Matsui M, Stoop CD, von Arnim AG, Wei N, Deng XW (1995) Arabidopsis COP1 protein specifically interacts in vitro with a cytoskeleton-associated protein, CIP1. Proc Natl Acad Sci USA 92:4239–4243

Mattaj IW, Englmeier L (1998) Nucleocytoplasmic transport: the soluble phase. Annu Rev Biochem 67:265–306

Meinkoth JL, Alberts AS, Went W, Fantozzi D, Taylor SS, Hagiwara M, Montminy M, Feramisco JR (1993) Signal transduction through the cAMP-dependent protein kinase. Mol Cell Biochem 127–128:179–186

Moll T, Tebb G, Surana U, Robitsch H, Nasmyth K (1991) The role of phosphorylation and the CDC28 protein kinase in cell cycle-regulated nuclear import of the *S. cerevisiae* transcription factor SWI5. Cell 66:743–758

Momand J, Wu HH, Dasgupta G (2000) MDM 2-master regulator of the p53 tumor suppressor protein. Gene 242:15–29

Moon RT, Brown JD, Torres M (1997) WNTs modulate cell fate and behavior during vertebrate development. Trends Genet 13:157–162

Moroianu J (1999) Nuclear import and export: transport factors, mechanisms and regulation. Crit Rev Eukaryot Gene Expr 9:89–106

Nasmyth K, Adolf G, Lydall D, Seddon A (1990) The identification of a second cell cycle control on the HO promoter in yeast: cell cycle regulation of SWI5 nuclear entry. Cell 62:631–647

Newlon MG, Roy M, Morikis D, Hausken ZE, Coghlan V, Scott JD, Jennings PA (1999) The molecular basis for protein kinase A anchoring revealed by solution NMR. Nature Struct Biol 6:222–227

Nigg EA, Hilz H, Eppenberger HM, Dutly F (1985) Rapid and reversible translocation of the catalytic subunit of cAMP-dependent protein kinase type II from the Golgi complex to the nucleus. EMBO J 4:2801–2806

Ohno M, Fornerod M, Mattaj IW (1998) Nucleocytoplasmic transport: the last 200 nanometers. Cell 92:327–336

O'Neill EM, Kaffman A, Jolly ER, O'Shea EK (1996) Regulation of PHO4 nuclear localization by the PHO80–PHO85 cyclin-CDK complex. Science 271:209–212

Pines J, Hunter T (1994) The differential localization of human cyclins A and B is due to a cytoplasmic retention signal in cyclin B. EMBO J 13:3772–3781

Raingeaud J, Gupta S, Rogers JS, Dickens M, Han J, Ulevitch RJ, Davis RJ (1995) Pro-inflammatory cytokines and environmental stress cause p38 mitogen-activated protein kinase activation by dual phosphorylation on tyrosine and threonine. J Biol Chem 270:7420–7426

Reiser V, Ammerer G, Ruis H (1999a) Nucleocytoplasmic traffic of MAP kinases. Gene Expr 7:247–254

Reiser V, Ruis H, Ammerer G (1999b) Kinase activity-dependent nuclear export opposes stress-induced nuclear accumulation and retention of Hog1 mitogen-activated protein kinase in the budding yeast *Saccharomyces cerevisiae*. Mol Biol Cell 10:1147–1161

Robinson MJ, Cobb MH (1997) Mitogen-activated protein kinase pathways. Curr Opin Cell Biol 9:180–186

Saavedra C, Tung KS, Amberg DC, Hopper AK, Cole CN (1996) Regulation of mRNA export in response to stress in *Saccharomyces cerevisiae*. Genes Dev 10:1608–1620

Saavedra CA, Hammell CM, Heath CV, Cole CN (1997) Yeast heat shock mRNAs are exported through a distinct pathway defined by Rip1p. Genes Dev 11:2845–2856

Schroeter EH, Kisslinger JA, Kopan R (1998) Notch-1 signalling requires ligand-induced proteolytic release of intracellular domain. Nature 393:382–386

Shiozaki K, Russell P (1996) Conjugation, meiosis, and the osmotic stress response are regulated by Spc1 kinase through Atf1 transcription factor in fission yeast. Genes Dev 10:2276–2288

Siebenlist U, Franzoso G, Brown K (1994) Structure, regulation and function of NF-kappaB. Annu Rev Cell Biol 10:405–455

Smith A, Ward MP, Garrett S (1998) Yeast PKA represses Msn2p/Msn4p-dependent gene expression to regulate growth, stress response and glycogen accumulation. EMBO J 17:3556–3564

Stacey MG, Hicks SN, von Arnim AG (1999) Discrete domains mediate the light-responsive nuclear and cytoplasmic localization of *Arabidopsis* COP1. Plant Cell 11:349–364

Stathopoulos-Gerontides A, Guo JJ, Cyert MS (1999) Yeast calcineurin regulates nuclear localization of the Crz1p transcription factor through dephosphorylation. Genes Dev 13:798–803

Stommel JM, Marchenko ND, Jimenez GS, Moll UM, Hope TJ, Wahl GM (1999) A leucine-rich nuclear export signal in the p53 tetramerization domain: regulation of subcellular localization and p53 activity by NES masking. EMBO J 18:1660–1672

Sträßer K, Hurt E (1999) Nuclear RNA export in yeast. FEBS Lett 452:77–81

Toda T, Cameron S, Sass P, Zoller M, Scott JD, McMullen B, Hurwitz M, Krebs EG, Wigler M (1987) Cloning and characterization of *BCY1*, a locus encoding a regulatory subunit of the cyclic AMP-dependent protein kinase in *Saccharomyces cerevisiae*. Mol Cell Biol 7:1371–1377

Toone WM, Jones N (1998) Stress-activated signalling pathways in yeast. Genes Cells 3:485–498

Treisman R (1996) Regulation of transcription by MAP kinase cascades. Curr Opin Cell Biol 8:205–215

Treitel MA, Kuchin S, Carlson M (1998) Snf1 protein kinase regulates phosphorylation of the Mig1 repressor in *Saccharomyces cerevisiae*. Mol Cell Biol 18:6273–6280

Turpin P, Hay RT, Dargemont C (1999) Characterization of IkappaBalpha nuclear import pathway. J Biol Chem 274:6804–6812

Uno I, Oshima T, Ishikawa T (1988) Localization of the regulatory subunit of cAMP-dependent protein kinase in *Saccharomyces cerevisiae*. Exp Cell Res 176:360–365

von Arnim AG, Deng XW (1994) Light inactivation of *Arabidopsis* photomorphogenic repressor COP1 involves a cell-specific regulation of its nucleocytoplasmic partitioning. Cell 79: 1035–1045

Wilkinson MG, Samuels M, Takeda T, Toone WM, Shieh JC, Toda T, Millar JB, Jones N (1996) The Atf1 transcription factor is a target for the Sty1 stress-activated MAP kinase pathway in fission yeast. Genes Dev 10:2289–2301

Wurgler-Murphy SM, Maeda T, Witten EA, Saito H (1997) Regulation of the *Saccharomyces cerevisiae* HOG1 mitogen-activated protein kinase by the PTP2 and PTP3 protein tyrosine phosphatases. Mol Cell Biol 17:1289–1297

Yamamoto N, Deng XW (1999) Protein nucleocytoplasmic transport and its light regulation in plants. Genes Cells 4:489–500

Yan C, Lee LH, Davis LI (1998) Crm1p mediates regulated nuclear export of a yeast AP-1-like transcription factor. EMBO J 17:7416–7429

Yang J, Bardes ES, Moore JD, Brennan J, Powers MA, Kornbluth S (1998a) Control of cyclin B1 localization through regulated binding of the nuclear export factor CRM1. Genes Dev 12: 2131–2143

Yang J, Drazba JA, Ferguson DG, Bond M (1998b) A-kinase anchoring protein 100 (AKAP100) is localized in multiple subcellular compartments in the adult rat heart. J Cell Biol 142:511–522

Yu Y, Jiang YW, Wellinger RJ, Carlson K, Roberts JM, Stillman DJ (1996) Mutations in the homologous *ZDS1* and *ZDS2* genes affect cell cycle progression. Mol Cell Biol 16:5254–5263

Zhu J, McKeon F (1999) NF-AT activation requires suppression of Crm1-dependent export by calcineurin. Nature 398:256–260

Subject Index

Printing (Computer to Film): Saladruck Berlin
Binding: Stürtz AG, Würzburg